The Fisheries Frontier

The Fisheries Frontier

Food Security, Quotas, and Conflict in Warming Seas

By

Maren Ostergaard

Vij Books

New Delhi (India)

Published by

Vij Books
(Publishers, Distributors & Importers)
4836/24, 3rd Floor, Ansari Road
Delhi – 110 002
Phone: 91-11-43596460
Mobile: 98110 94883
e-mail: contact@vijpublishing.com
www.vijbooks.in

ISBN: 978-93-7794-005-8 (HB)

ISBN: 978-93-7794-006-5 (PB)

ISBN: 978-93-7794-007-2 (eBook)

This book was developed using advanced writing tools to support the drafting and ideation process. All content has been thoughtfully curated, refined, and reviewed to ensure quality and relevance.

Contents

Introduction

On 3 October 2018, in Ilulissat, Greenland, diplomats from ten parties signed the Agreement to Prevent Unregulated High Seas Fisheries in the Central Arctic Ocean. Outside, the air had the metallic bite that belongs to ice and salt; inside, pens scratched across paper for a treaty built around an absence. There was, at that moment, no commercial fishery in the central Arctic high seas to stop. The signatories were not dividing a proven abundance. They were trying to get ahead of a moving problem by writing rules before the seam opened. That seam matters: the line where ecology outruns law, where a warming sea produces not instant chaos but a slower, more exacting political strain.

The map stayed still; fish moved.

That is the Arctic difficulty in miniature. States inherit coordinates, licences, patrol districts, legal doctrines and habits of diplomatic speech that all presume a certain stability in where value will appear. Fish do not respect any of it. They follow temperature gradients, prey fields, currents, spawning grounds, and seasonal timing, which can shift unevenly and sometimes suddenly. A cod stock does not cross a maritime boundary as an argument; it crosses as a biological response. Yet the moment its distribution changes, the argument begins, because catch histories, quota formulas, processing investments and regional livelihoods were built on an older geography. Scarcity alone does not explain the friction that follows. More often, trouble gathers when institutions are calibrated to yesterday's distribution and fleets are already fishing tomorrows.

The Arctic sharpens this tension because distance is not a backdrop there; it is an active force in governance. Long transits consume fuel and patience. Darkness obscures routine verification. Weather narrows the window in which a patrol vessel can board safely, in which an aircraft can fly, in which an inspector can tell a paperwork discrepancy from a deliberate falsehood. Even the most ordinary decisions become expensive. A port inspection delayed by storms is not merely inconvenient. A missed satellite pass, an overworked analyst, a camera unit that fails in freezing spray, a boarding postponed until morning—such small frictions accumulate. They shape what compliance means in practice, because rules that look precise on paper may prove blunt in distant seas. In the Arctic, enforcement is rarely a matter of constant presence. It is more often a matter of selective visibility.

That phrase matters because fisheries conflict usually begins well before gunboats, headlines or overt crisis. It begins when the visible and the verifiable start to drift apart. A vessel slows near a boundary and insists its gear was deployed legally. An allocation formula still privileges historical catch, even though the stock's seasonal centre of gravity has shifted. A port inspector sees discrepancies in a landing declaration but lacks the chain of evidence to prove deliberate misreporting. A minister, pressed by domestic processors and coastal constituencies, hardens a line that scientists have already complicated. None of these events needs to be dramatic to matter. The danger lies in their combination. A system that cannot compare what is claimed with what is done, and cannot revise bargains without looking politically weak, becomes brittle even when the fishery remains biologically productive.

This is why warming seas do not produce a simple story of less fish, more conflict. In some places, changing conditions may create new opportunities, at least for a time. Access can improve for one state even as it deteriorates for another. New entrants can appear where older participants assumed precedence. One fleet may encounter a stock first; another may still possess the capital, processing capacity and legal standing to capture most of its value. Distributional conflict follows from these mismatches. The language of conservation, fairness, precaution and sovereign right then acquires a double life. It describes genuine concerns. It also becomes a bargaining vocabulary through which actors defend revenue streams, employment, food supply, regional identity and room for manoeuvre. The politics of fish are rarely just about fish.

Consider the northeast Atlantic mackerel dispute that sharpened after 2007, when the stock's summer distribution shifted markedly towards Icelandic and Faroese waters. By 2010 and 2011, Reykjavík and Tórshavn argued that quota shares rooted in earlier distributions had become indefensible. The European Union and Norway, whose fleets and processors had long benefited from the established arrangement, did not see the matter as a simple ecological correction. Negotiations turned acrid. Catch limits exceeded scientific advice. Trade measures followed, including European action against the Faroe Islands in 2013 before a settlement was reached in 2014. The scene looked, from a distance, like a disagreement over numbers. It was more than that. The concept at stake was allocation under moving baselines. Once a stock changes where it can be caught, every prior investment becomes evidence for someone: evidence of earned entitlement, of historical dependence, of unjust exclusion, of belated recognition. Ecological motion enters politics as a claim about who should bear adjustment costs.

The mackerel case also reveals a harder truth about scientific authority. Science can narrow uncertainty, but it does not dissolve interests. Survey coverage changes. Models inherit assumptions from older productivity patterns. Seasonal presence is not the same as annual attachment. If the question is whether fish now spend more time in one zone than before, data may help. If the question is how that fact should alter rights, data alone cannot answer it. That is where frustration begins. Policymakers often invoke better science as though it were a neutral solvent for conflict. Often it is indispensable. It is not sovereign over distributional politics. In warming seas, arguments about fish become arguments about baselines, and baseline disputes are political even when wrapped in technical prose.

The Arctic introduces another complication: not every gain in access is a gain in control. A stock moving northward may enter waters where monitoring is thinner, seasons are more compressed and mixed catches are harder to document. Pulse fisheries form under precisely these conditions. When fish arrive in dense, short-lived concentrations, speed becomes an advantage. So does ambiguity. A regulator may design a sound annual regime and still fail during the two weeks that matter most. The practical question is not whether a state has rules. It is whether it can test, compare and act quickly enough when incentives peak. Risk is visible here not as an abstract condition but as behaviour: more vessels clustered along boundaries, more logbook disputes, more pressure on observers, more landings routed through the few ports able to absorb them.

And yet the Arctic is not simply an arena of inevitable breakdown. One of the most important counter-examples lies in the Barents Sea, where Norway and Russia, through the Joint Norwegian-Russian Fisheries Commission established in 1976, have managed major shared stocks for decades under conditions that combine geopolitical sensitivity with ecological volatility. Their co-operation has not been pristine, and it has hardly floated above wider politics. Still, it has often produced a sturdier reality than the language surrounding it would suggest: joint scientific work, negotiated catch levels, technical measures, and enough institutional routine to keep disputes from turning into permanent crises. The point is not to romanticise the arrangement, still less to assume it can be copied wholesale elsewhere. The point is that durable management in difficult waters depends less on goodwill than on repeated procedures that can absorb disagreement before it hardens.

That distinction between goodwill and procedure is easy to miss, perhaps because fisheries governance still attracts two opposite simplifications. One treats conflict as the natural consequence of scarcity in a world of declining resources. The other assumes that once the right legal forum, treaty text or tracking technology is in place, behaviour will settle obediently around it. Both are too neat. Conflict is often generated not by absolute shortage but by lag: lag between stock movement and quota revision, between rule adoption and enforcement capacity, between detection and proof, between political rhetoric and the quieter needs of de-escalation. Likewise, technology does not rescue governance by itself. Automatic Identification System data can reveal suspicious gaps, but a gap is not a conviction. Cameras can document discards, but only if their footage is trusted, stored, reviewed and linked to sanctions that mean something. Satellites can narrow the search problem while leaving the attribution problem intact.

A fishing vessel is never only a fishing vessel in such circumstances. It is also a carrier of permissions: quota rights, fuel contracts, insurance cover, access to landing points, market relationships, documentation trails, flag-state credibility and, not least, the expectation that if challenged it can explain itself persuasively enough to avoid the costliest consequences. This is why ports matter so much. Port state measures appear administratively mundane beside the drama of patrols in rough seas, yet they often bite harder. Denial of landing, refuelling or services can change incentives at lower cost than a permanent offshore presence. Markets matter for the same reason. Traceability schemes, retailer requirements, financing conditions and insurance clauses distribute enforcement across institutions that are not patrol agencies but still determine who gets to operate. If a fisheries frontier has a hidden architecture, much of it sits onshore.

Still, onshore tools have their own politics. A traceability regime that reassures importers may burden small operators. A tougher inspection system may redirect trade into less transparent channels rather than stop suspect activity. Community legitimacy can fray when policies appear to protect distant reputations while ignoring local livelihoods. Food security enters here in a form more granular than a national slogan. There is a difference between preserving export earnings, preserving affordable domestic protein, and preserving the right of a small coastal or Indigenous community to maintain fishing as a source of sustenance and a social practice. Governments often compress these under a single banner of national interest. They are not the same. When quota systems concentrate ownership, when leasing prices rise beyond local reach, when adaptation demands capital that peripheral regions lack, grievance accumulates within states as well as between them. External postures can harden partly because internal bargains have already failed.

The seam appears again here. It is not only a line on a chart. It is the place where a statistical adjustment becomes a question of justice, where a monitoring device becomes a test of legitimacy, where a boarding decision becomes a message. In distant seas, even routine enforcement carries an element of signalling because every intervention can be interpreted twice: once as administration, once as stance. That ambiguity is especially useful to operators willing to test limits. Grey behaviour thrives in such spaces. It is not always flamboyantly illegal. Often it is merely evasive, calibrated, opportunistic. A vessel fishes hard against a boundary. A company exploits layered ownership to blur responsibility. A captain complies slowly enough to extract advantage while preserving a legal defence. These are not anomalies at the edge of governance. They are part of the incentive structure produced when rules are valuable, visibility is partial and sanction pipelines are uneven.

The central Arctic fisheries agreement signed in Ilulissat in 2018 makes more sense when seen in this light. It was not only an act of precaution about a fishery that might emerge. It was a recognition that waiting for commercial pressure to arrive would make every later decision harder. Better to establish research obligations, consultation procedures and a norm against unregulated exploitation before the seam widened further. Yet that achievement also reveals a limitation. Pre-emption works best where interests are still partially hypothetical. Once fleets, ports and domestic constituencies are deeply invested, flexibility narrows. The hard cases arrive when institutions must revise existing entitlements under stress, not merely design future ones in calmer conditions.

That is why the most useful questions are diagnostic rather than theatrical. Not whether conflict is coming as if from the weather, but where capacity is thin, where incentives are sharpest, where visibility collapses, where procedural exits still exist. Can the relevant parties measure a distribution shift in a way others accept? Can they translate that measurement into a revisable allocation without inviting political humiliation? Can suspicious conduct be attributed to a vessel, an owner and a chain of landings strongly enough to sustain a sanction? Can a detention, seizure or inspection dispute be channelled into routine diplomacy before cameras and nationalist rhetoric turn it into a test of resolve? Stability in the Arctic will not mean the end of disagreement. It will mean rules credible enough under change that disagreement does not automatically become confrontation.

There is no shortage of grand language about the Far North. Much of it mistakes remoteness for emptiness. The Arctic is not empty. It is crowded with claims, methods, memories of earlier bargains, expensive infrastructures, fragile communities, bureaucratic seams and increasingly mobile life beneath the surface. The practical challenge is not to freeze that movement in place. It is to govern motion without pretending that sovereignty alone can do the work, or that science alone can settle what politics has to distribute. Between the patrol vessel and the processing plant, between the stock assessment and the parliamentary speech, between the vessel track and the evidentiary file, a great deal of conflict is made or unmade.

The decision that shadows all of this is not whether to choose order over disorder. That is too easy. It is whether institutions can be designed to revise access, verify conduct and preserve legitimacy at the same time, in conditions that punish delay and reward ingenuity. Every fisheries dispute in warming seas asks some version of that question. The answer will rarely arrive in a single treaty text or a single enforcement action. It will emerge from how states, fleets, ports, markets and communities handle the next shift before it hardens into precedent. The Arctic seam is already opening. What matters now is whether it becomes a fault line.

Chapter 1

Arctic Fisheries as Frontier

How does a state govern a fish stock that no longer behaves like the map? That question sits at the heart of the Arctic frontier, where political lines remain fixed while cold-water life shifts with currents, temperature, prey, and season. The contradiction is stark. A boundary can be surveyed, legislated, and defended. A fish stock can do none of those things. It moves when conditions tell it to move, and every such movement tests the settlement built around where it used to be.

The Arctic makes that test unusually severe because nature does not merely complicate management there; it disciplines it. Distance lengthens every response. Ice narrows access. Darkness conceals ordinary behaviour until it begins to look deliberate. A vessel in a temperate sea may be inspected as a matter of routine. A vessel in far northern waters may be inspected only after costly travel, uncertain weather, and a series of risk-related choices. Cold gets inside everything. It slows machines, people, judgement, and proof. Even the paper trail can arrive late.

That physical severity is easy to romanticise from afar. It should not be. Harsh operating conditions do not create moral clarity; they create administrative limits. Rules that appear crisp in an office can fray at sea when enforcement depends on timing, fuel, visibility, and safe boarding conditions. A regulator may know what ought to happen and still fail to make it happen at the required moment. A fleet may comply in general and still discover that the profitable edge lies in the gaps between intention and reach. The frontier begins there, not as myth but as mismatch.

The deeper strain follows from movement. Fish do not migrate into a political vacuum. They move through systems of entitlement, expectation, and memory. A quota share reflects older patterns of access. A processing plant reflects past certainty about where landings will arrive. A coastal community reflects years of investment made on the assumption that a season would keep roughly the same shape. When a stock shifts, the catch does not simply appear somewhere else; it reorders who sees value first, who bears more risk, and who suddenly argues that the old bargain no longer describes reality. What looked settled starts to feel provisional.

That is why Arctic fisheries deserve to be treated as frontier politics rather than as a narrow technical problem. Frontier does not mean lawless. It means unstable in a specific way: institutions remain present, but the facts they were built to govern begin to slip. The state still issues licences, patrols waters, prosecutes breaches, and speaks in the language of jurisdiction. Fleets still plan routes, secure finance, hire crews, and chase margins. Yet beneath this formal continuity, the object of governance is in motion. That motion turns management into distributional conflict, because the central argument is no longer only how much may be taken, but who may take it, where, when, and with what claim to fairness.

The stakes widen quickly. Fish can be food, wage, export, symbol, and bargaining chip all at once, but not in equal measure for every actor. One government worries about affordable supply and domestic stability. Another worry is about quota value, vessel viability, and the pressure exerted by processors, ports, and regional employers. One community sees fishing as a daily livelihood and an inherited right. Another sees it as revenue, throughput, and industrial continuity. These are not semantic differences. They produce different tolerances for compromise, delay, and uncertainty. They also explain why fish can acquire strategic weight without becoming a military question in any simple sense. The contest often lies in access, verification, and political room to yield.

Still, not every disagreement in northern waters should be inflated into a coming crisis. Co-operation persists, often quietly, and much of fisheries governance depends on procedures so ordinary that they attract little attention when they work. Meetings, inspections, data exchanges, licensing routines, technical arguments over evidence, and practical co-ordination between agencies can hold a great deal of strain without spectacle. And yet their quietness is exactly what makes them vulnerable to neglect. When the public notices fisheries, it is usually because something has gone wrong: a detention, an accusation, a stock under pressure, a boundary incident, a claim that someone else is taking what should not be taken. By then, the administrative groundwork has often already weakened.

A more useful way to think about these waters begins with three linked realities. Governance depends on capacity: not merely formal authority, but the practical ability to monitor, compare, verify, and act at a distance.

Behaviour depends on incentives: not abstract motives, but the concrete rewards attached to early access, quota share, market timing, and regulatory ambiguity. Order depends on visibility: what can be seen, what can be recorded, what can be trusted, and what can be proved when claims begin to clash. Read together, those realities make the Arctic legible without flattening its complexity. They keep attention fixed on institutions and conduct rather than on melodrama.

The path ahead follows the shape of the problem itself. It begins with the hard material facts of geography, distance, and seasonality, because ambition means little if the operating environment defeats it. From there the argument moves to the friction between static jurisdiction and mobile resources, where shifting distributions transform routine management into a contest over allocation and precedence. The pressure then broadens into food, employment, and strategic value before settling on the practical lens needed to judge what states and fleets actually do when rules meet moving fish and imperfect sight.

Everything turns, in the end, on a stubborn question. When the sea changes faster than the bargain, what counts as effective control: the rule on paper, the patrol that may never arrive, the landing that can still be denied, or the shared restraint that prevents the contest from hardening? The Arctic does not permit easy answers. It asks for something harder - a clear look at what distance, motion, and limited visibility do to power before the first boat is even boarded.

Geography, distance, and seasonality

In October 2018, representatives of ten parties met in Ilulissat, Greenland, to sign the Agreement to Prevent Unregulated High Seas Fisheries in the Central Arctic Ocean. The immediate result was oddly negative: they agreed to restrain a commercial fishery that did not yet exist. That choice made sense only if one began with geography rather than optimism. The central Arctic is not merely cold water on a map. It is water made difficult by reach, by ice, by darkness, by sparse infrastructure, and by the blunt fact that once something goes wrong, help often starts far away. Distance changes behaviour. It also changes what a rule can accomplish.

The point seems obvious until policy language starts to float free of the sea itself. From an office, enforcement can look like a matter of statutes, coordination, and will. At sea, especially in high latitudes, each of those abstractions acquires weight. A patrol aircraft needs the weather in which it can fly. An inspector needs a vessel that can get there. A boarding team needs light, sea room, and a captain prepared to co-operate before conditions deteriorate. A fishing vessel needs fuel, ice clearance, crew endurance, and enough certainty about the trip to justify the cost. Geography is not scenery. It is the first regulator.

That is why the Arctic frontier begins with operating conditions rather than with law or diplomacy. Ice edge dynamics alter where vessels can safely work and where patrols can plausibly follow. Freeze-up and break-up change the timing of access. Polar night shortens the hours in which visual identification, gear inspection, and small-boat operations remain straightforward. Long transits compress decision time because there is less room to be wrong. If a suspicious contact appears near a boundary in a temperate sea, authorities may still have options after a delay. In remote northern waters, delay can decide the matter. By the time a patrol arrives, the vessel may have crossed a line, changed its behaviour, transferred catch, or reached conditions under which boarding no longer makes sense. The sea closes the interval in which law can act.

This compression matters because fisheries governance depends on sequences rather than isolated acts. A vessel is detected, identified, tracked, queried, approached, boarded, documented, and, if necessary, sanctioned. Each step can fail for ordinary reasons. Bad weather obscures radar returns. Ice or swell makes transferring between vessels unsafe. Darkness turns a visible deck operation into a matter of inference. Communications lag. Crews tire. The result is not theatrical lawlessness but something harder to manage: a steady widening gap between what authorities nominally control and what they can verify in real time. The name for that gap is not weakness alone. It is friction. And friction is political, because it sets the outer limit of any promise a state can credibly make.

The Barents Sea offers one of the clearest illustrations. Since 1976, Norway and the Soviet Union, and later Russia, have managed shared fish stocks through the Joint Norwegian-Russian Fisheries Commission. The arrangement is often cited because it has endured, but its durability rests partly on an unromantic adjustment to geography. These are productive northern waters, yet they are still defined by long steaming distances, dangerous winter conditions, and the need to combine scientific assessment with practical control. No serious authority in that region has governed on the assumption of omnipresence. Instead, the system has relied on layers: agreed measures, reporting rules, coast guard activity, technical co-operation, and port-based follow-through. That is a physical lesson before it is an institutional one. Where sea conditions punish constant interception, enforcement shifts towards selective presence, routine documentation, and choke points on land.

The same logic helps explain why the central Arctic agreement in Ilulissat took a precautionary form. The governing problem was not only uncertainty about future stock distributions. It was uncertainty under extreme operating conditions. If a fishery were to emerge in waters where patrols would be costly, seasonal, and politically delicate, waiting for competition to begin would load every later disagreement with higher stakes. Pre-emption was therefore less a triumph of foresight in the abstract than an acknowledgement of geography's tax on response. Once fleets, processors, and domestic constituencies invest, the sea becomes harder to govern because every intervention arrives later and costs more.

Cost itself deserves more attention than it usually receives. "Remote" can sound like a descriptive term. In practice it is a budget line. Longer voyages raise fuel consumption, maintenance needs, insurance burdens, and the opportunity cost of tying up patrol assets for days or weeks. For fishing operators, these expenses reward concentration. They make short productive windows more valuable and can encourage behaviour that pushes against boundaries, seasons, and reporting obligations. For inspectors, the same expenses force prioritisation. One area is covered; another is not. One vessel is checked; another is not. A surveillance system can detect anomalies across a vast area, but only a fraction of them can be pursued. Efficiency, in this setting, is not a managerial ideal. It is a rationing problem. Someone must decide which uncertainty to live with.

Seasonality sharpens every part of that calculation. Many Arctic fisheries and survey efforts operate in windows set by ice retreat, stock movement, weather, and light. These windows do not simply organise work; they concentrate pressure. When access opens, vessels move quickly because delay can mean missing the fish, the market, or the safe return passage. Regulators face the mirror image of that urgency. Inspections cannot be distributed evenly across the year if the relevant activity is densely packed into a narrower period. A regime that looks adequate on an annual chart may fail in the weeks that matter most. This is one reason "effective enforcement" in the Arctic cannot mean the same thing it might mean nearer dense ports and shorter routes. The benchmark changes. Success may lie less in constant interception than in credible unpredictability, in reliable paperwork, in data that can later support sanction, and in port controls that make sea-based evasion less profitable.

Light, or its absence, deserves its own seriousness. Polar darkness is not a poetic hardship. It alters what can be seen, by whom, and with what confidence. Human observers work differently at night. Cameras work differently. Aircraft operations become more restrictive. Distinguishing one vessel from another, reading markings, confirming gear configuration, or documenting an interaction between vessels all become harder. The same is true in heavy fog, icing conditions, or rough seas, but the Arctic compounds these difficulties over long periods. Visibility then becomes a legal issue, not merely a sensory one. If an infraction cannot be clearly attributed, penalties weaken. If penalties weaken, deterrence thins. The path from darkness to dispute is shorter than it appears.

This is where geography and behaviour meet. A fleet that knows authorities cannot watch everywhere may not need to violate rules openly. It may only need to exploit the edge of uncertainty. Fishing near a line rather than across it. Delaying compliance with an instruction until the weather worsens. Using timing, route choice, or transhipment patterns that remain technically arguable while frustrating clear proof. The physical setting does not cause such conduct in any mechanical sense, but it rewards operators who understand how conditions erode verification. Risk becomes visible here not as an abstract probability but as a choice: whether to steam farther for a check, whether to board in worsening seas, whether to trust a declaration when a hold cannot be safely inspected, whether to escalate a doubtful case with incomplete evidence.

A useful limitation must be admitted. Harsh conditions do not make governance impossible, nor does temperate geography guarantee control. The North Atlantic has long produced disputes, evasions, and dangerous fisheries politics despite shorter distances and denser institutional coverage. Illegal, unreported, or disputed activity can flourish close to shore if incentives are strong and penalties are weak. Arctic severity explains a great deal, but not everything. The distinction matters because it prevents a lazy determinism. Geography sets a ceiling for policy ambition; it does not write the entire script. Institutions still matter, and so do markets, diplomatic routines, and domestic political choices.

Even so, the Arctic ceiling is lower than many official narratives imply. A state may possess a legal right and still lack the practical means to exercise it quickly or continuously over distant waters. A surveillance system may gather abundant data and still fail at the decisive moment because analysts cannot fuse it fast enough, because weather blocks response, or because proof gathered under difficult conditions collapses under legal scrutiny.

Administrative reach becomes patchy at the very moment when ecological movement and concentrated fishing effort demand precision. That is why northern fisheries governance depends so heavily on calibration. Rules must be written for the sea that exists, not for the map imagined in the capital.

The image that returns, again and again, is the long transit. It is a mundane phrase, almost bureaucratic, yet it captures the governing fact of the Arctic better than grander language does. A long transit means slower response, dearer presence, narrower margins for error, and more discretion delegated to whoever is already on scene. It means the weather has more time to change the argument. It means evidence can degrade before it is collected. It means that routine enforcement decisions acquire diplomatic weight because each intervention is costly enough to look intentional. In calmer and closer waters, a boarding can remain administrative. Far north, the same act can read as policy.

The physical setting therefore does more than make governance expensive. It shapes the very definition of what states can plausibly promise, what fleets expect they can test, and what kinds of order can endure without constant spectacle. If the next suspicious track appears near the ice edge at the end of a long transit, under failing light and in worsening seas, what exactly counts as control: the capacity to arrive, the capacity to prove, or the capacity to make that voyage unnecessary through rules designed for distance rather than embarrassed by it?

From fixed lines to moving resources

In 2010, Iceland set a unilateral mackerel quota after the stock's summer distribution had expanded markedly into Icelandic waters, and a dispute that had been brewing in the northeast Atlantic hardened into a test of political arithmetic. The fish had moved. The bargain had not. What followed was not a collapse of law, nor a sudden absence of institutions, but a more awkward and more revealing spectacle: coastal states arguing over whether an established allocation still described the resource it purported to govern. The map remained crisp. The stock did not.

That contrast matters because fisheries management depends on a quiet fiction. We draw lines around water as though mobile life will remain sufficiently faithful to those lines for long enough to make bargains durable. Sometimes that fiction works well enough. A stock spends roughly the same seasons in roughly the same places; catches follow a recognisable pattern; quota shares, however contested, acquire the authority of habit.

But when distribution shifts, even modestly, routine management can become a dispute about entitlement. Fish are not only being caught. They are being counted as evidence. Presence in one zone becomes an argument for a larger share. Absence from another becomes a grievance. Mobility turns biology into allocation politics.

The key fact is simple. Jurisdiction is fixed faster than ecology. An exclusive economic zone can be charted, legislated, defended, and reproduced in licensing systems, satellite maps, and bureaucratic practice. A fish stock responds instead to temperature, prey, spawning conditions, currents, depth, and seasonality. Those movements may be gradual, abrupt, cyclical, or uneven across age classes. They may shift the centre of distribution, the timing of arrival, the density of local concentrations, or the route by which the stock becomes accessible to particular fleets. None of this need imply crisis by itself. Trouble begins when a management formula assumes that yesterday's spatial pattern still organises today's value.

The language used to describe such disputes can sound deceptively technical. States argue over zonal attachment, historical catches, access rights, baseline periods, and scientific indicators. Each phrase appears tidy. Each carries a distributive punch. Historical catch rewards those who secured access under an earlier geography. Zonal attachment gives weight to where the fish are found, but that apparently plain idea opens a nest of complications. Which season counts? Which age classes? Which depth range? Which measure of "presence" matters most: biomass, duration, or commercial accessibility? Once the stock moves, every metric becomes a site of negotiation. Measurement itself acquires politics.

The northeast Atlantic mackerel dispute remains instructive precisely because it was not an Arctic morality play staged at the edge of the world, but a recognisable case of moving resources straining inherited bargains. By the late 2000s, Iceland and the Faroe Islands argued that a stock long managed chiefly by the European Union and Norway now spent enough time in their waters to justify a far larger share of the total catch. The established parties did not reject ecology altogether; they rejected what they saw as opportunistic arithmetic and destabilising unilateralism. Scientific advice was available, but advice on total catch could not by itself resolve the prior political question of who should receive what share. The conflict then escaped the narrow frame of stock assessment. It became a quarrel about fairness, discipline, and the price of reopening a bargain under pressure. Fish moved north and west. So did the centre of the argument.

A demersal stock makes the frontier logic even clearer because the shift is often less theatrical than with pelagic fish and more consequential for those who invested in fixed expectations. Consider cod in the Barents Sea and adjacent northern waters. Cod does not drift as an abstract biomass through diplomatic prose. It occupies habitats, follows feeding opportunities, responds to temperature, and becomes commercially visible in patterns that matter to fleets with different steaming distances, gear, ports, and seasonal timing. If a greater share of the stock becomes accessible earlier in one zone than another, that is not a mere ecological curiosity. It changes who catches first, who fills quotas sooner, who lands more fish into domestic processors, and who then claims that formal allocation no longer tracks operational reality. The question is not only where the stock exists. It is where it becomes usable, and by whom.

Catch it first, claim it harder.

That terse frontier rule does not describe every fishery, but it captures a recurrent temptation. Access creates precedent, and precedent hardens quickly when vessels, plants, financiers, and local politicians begin to treat a recent pattern as the obvious one. A resource that appears in commercially viable concentrations inside one jurisdiction can transform the internal politics of that state before any formal renegotiation occurs. Ports plan around new landings. Processors hire on the assumption of continuity. Ministers absorb the language of "our fish" from regions that had little reason to use it a decade earlier. When negotiations resume, they are no longer arguments over a hypothetical distribution shift. They are arguments shadowed by sunk costs and sharpened constituencies. Mobility has done more than move biomass. It has manufactured domestic pressure.

This is why disputes over moving stocks often become more bitter than disputes over declining stocks. A shrinking resource at least presents an obvious common problem, even if states still quarrel over burden-sharing. A shifting resource creates asymmetry. One side experiences loss. Another experiences arrival. The gaining side speaks the language of correction: the stock is here, the quota should follow. The losing side speaks the language of order: management cannot survive if every fluctuation triggers a new division. Both claims can be sincere. Both can also conceal self-interest. What matters is that ecological change does not arrive in politics as neutral information. It arrives as leverage in argument, though not the mechanical sort beloved of policy jargon. It arrives as landed weight, altered routes, changed timing, and fresh demands from people who can point to fish in local waters and ask why law is lagging behind the sonar.

The legal structure of maritime governance intensifies this friction because it was built to stabilise space, not to keep pace with restless distributions. The modern law of the sea gives coastal states extensive rights within their exclusive economic zones and leaves the high seas subject to more collective forms of management. That arrangement is coherent as a spatial settlement. It becomes awkward when stocks move across those spatial categories in ways that unsettle earlier assumptions. A fish once caught chiefly within national zones may spend more time near a boundary, pass seasonally through several zones, or extend into adjacent high seas pockets where monitoring is weaker and incentives to race are stronger. The line on the chart remains exact. The resource relationship to that line does not. The seam opens.

A seam is not yet a rupture. The distinction matters. Many shared stocks move across jurisdictions without producing diplomatic theatre, because institutions have already created routines for adjustment, data exchange, and joint decision-making. The Joint Norwegian-Russian Fisheries Commission offers an important, if not flawless, example. Since 1976, it has provided a structure through which major Barents Sea stocks have been assessed and managed cooperatively despite wider geopolitical strain. That record should not be romanticised, but neither should it be dismissed. It shows that mobility need not destroy order if parties accept that the relevant unit of governance is the stock's behaviour rather than the administrative convenience of the boundary. The limitation is equally clear. Such arrangements depend on political discipline, scientific working relationships, and enough trust in process to keep disagreement from becoming proof of betrayal. They are hard-won. They do not appear automatically when fish shift.

And yet even strong cooperative structures face a problem they cannot fully solve: the difference between movement and meaning. A stock can move without states agreeing on what that movement should count for. One party may treat a temporary seasonal increase in local abundance as evidence of a structural shift. Another may insist that spawning grounds, recruitment patterns, or long-run distribution still justify the older allocation. Both can marshal scientific language. Neither is merely "anti-science". The dispute lies in the translation from ecological fact to political entitlement. This is why more data, though necessary, often fails to settle the decisive question. Data can narrow uncertainty about distribution. It cannot determine how adjustment costs should be allocated among parties with different dependencies and different memories of prior sacrifice.

That translation problem becomes especially acute when resources move into high seas pockets or other regulatory seams. Within a national zone, a coastal state can at least assert a clear legal basis for licensing, inspection, and sanction. Beyond those boundaries, governance often relies on regional fisheries bodies, negotiated measures, flag-state performance, and the hope that transparency will constrain opportunism. When a commercially valuable stock spends meaningful time in such spaces, the pressure to exploit ambiguity rises. Fleets may race to establish a presence. States may harden positions to avoid setting precedents that weaken future claims. The issue then is not simply access. It is the fear that hesitation today will be counted as surrender tomorrow. Mobility converts prudence into perceived loss.

One reason this dynamic proves so stubborn is that fixed lines serve many functions at once. They are legal boundaries, administrative shortcuts, fiscal assumptions, and emotional anchors. Communities invest in the idea that a fishery belongs, at least partly, to a recognisable space. Bureaucracies prefer distributions that fit their filing systems. Politicians prefer maps that reassure voters. Fish, by contrast, remain indifferent to every one of those preferences. The encounter between line and life is therefore not just technical. It is psychological. A stock that moves can make an entire governance system feel briefly unreal, because rules that seemed settled now depend on persuading others that the old geography still deserves obedience.

The more serious response is neither to fetishise fluidity nor to defend rigidity for its own sake. It is to recognise that mobility changes the nature of the bargain. In a stable fishery, allocation may look like a one-off settlement followed by periodic technical maintenance. In a moving fishery, allocation becomes a continuing argument over timing, evidence, and revision. That does not mean perpetual renegotiation. Constant reopening can itself destroy discipline. But it does mean that any durable arrangement must decide, sooner rather than later, what kind of movement matters, how it will be measured, over what period, and with what consequences for access. Without such terms, every ecological shift risks being litigated through unilateral quotas, diplomatic threats, and increasingly moralised language about fairness.

The frontier quality of Arctic fisheries lies precisely here. Frontiers are not defined only by remoteness; they are defined by the instability of the relationship between authority and value. A state may still control a zone while losing confidence that the zone contains the resource in the pattern assumed by its institutions. Another may gain access without yet possessing the legitimacy or infrastructure to convert that access into accepted right. The line remains. The resource moves. Politics fills the gap. Once that is understood, the argument over fisheries looks less like a technical dispute interrupted by occasional drama and more like a continuous effort to keep ecological motion from becoming distributive shock.

The quiet test is harder than it sounds. When a stock shifts poleward or spends longer in a different zone, what exactly deserves protection: the historical bargain that enabled long investment, the present distribution that makes exclusion look absurd, or the procedural restraint that prevents every new season from becoming a fresh quarrel over ownership? The answer cannot be read off the chart, however clean the line.

Food, jobs, and strategic value

In August 2013, the European Union imposed measures against the Faroe Islands over the dispute surrounding Atlanto-Scandian herring and north-east Atlantic mackerel. The outcome was concrete and unmistakably economic: Faroese herring and mackerel faced restrictions in the European market, and certain Faroese vessels were denied access to EU ports. No naval confrontation followed. No one needed one. A stock allocation dispute had already crossed into strategic territory because fish were never just biomass in the water. They were export earnings, factory throughput, crew wages, and political standing. A crate on the quay can carry more argument than a speech.

That is the first distinction that serious fisheries politics requires. "Food security" and "the fishing economy" are often spoken in one breath, as though they refer to the same interest viewed from different angles. They do not. A government worried about domestic food supply asks one set of questions: will households have access to affordable protein, will local markets remain supplied, will interruptions in imports or local catches produce visible social strain? A government worried about export revenue asks another: what will happen to foreign exchange earnings, tax receipts, processing margins, and the commercial viability of fleets built for external sales? Add employment, and the picture changes again. A state may tolerate higher consumer prices while fighting hard to preserve jobs in remote ports, freezing plants, repair yards, and transport chains. Same fish, different pressure.

The difference matters because political language often conceals it. Ministers invoke "national interest" when they mean several interests at once, some of them aligned and some not. A coastal town that depends on landings wants continuity of work. A treasury wants revenue and investment confidence. An urban consumer wants stable prices. A processor wants predictable volume. A quota holder may want maximum asset value even if the fish are landed elsewhere. None of these goals is illegitimate. They simply do not point in the same direction when stocks move or access is constrained. Once that divergence appears, negotiation hardens. What looks from the outside like stubbornness may be a government's attempt to hold together constituencies whose priorities no longer match.

The Faroese dispute exposed this problem precisely. Brussels treated the matter as one of quota discipline and collective management. Torshavn treated it as a dispute about fair access to fish that had become more available in Faroese waters. Both sides used the language of sustainability, but the intensity of the disagreement came from distribution and dependency. Quotas determine catches; catches determine landings; landings determine whether a factory line runs, whether crews sail, whether local lending decisions still make sense. When the European Union chose market restrictions rather than maritime force, it demonstrated something essential about fisheries strategy. The decisive instrument was not a gunboat. It was access to a buyer. Strategic value, in this domain, often moves through customs paperwork, sanitary clearance, and port entry rather than through overt coercion at sea.

That same logic helps explain why coastal states can behave very differently when faced with similar ecological change. Where fish are central to daily diet and local availability, governments may privilege supply and price stability. Where fisheries are strongly export-oriented, the emphasis shifts towards maintaining volumes, preserving reputation in foreign markets, and defending bargaining positions that protect future quota share. Where employment in peripheral regions carries special political weight, leaders may fight hardest not for the highest theoretical national return but for landings that keep specific communities alive. The abstraction "risk" only becomes real once it is attached to a decision: whether to open a season early, whether to press for a larger share at a regional forum, whether to subsidise fleet renewal, whether to accept a compromise that secures conservation but redirects fish away from a domestic plant.

Iceland offers a sharp illustration of how these pressures can combine without becoming identical. When Iceland adopted unilateral mackerel quotas from 2010 onward, the argument was not simply that fish had entered Icelandic waters and should therefore be counted. The deeper claim concerned fairness to a national fishing economy that saw the established allocation as lagging behind ecological reality. That economy included vessels, crews, processors, and the politics of a country for which marine resources have long carried unusual weight. Yet Iceland's position also showed the distinction between food and commercial value. Mackerel was not politically central because Icelandic households would go hungry without it. It mattered because access to a shifting stock had become economically meaningful and symbolically charged. The fish had acquired strategic importance through allocation, not through any immediate food emergency.

This helps clear away a persistent confusion. Fish become strategic when control over access, processing, and trade affects a state's room for manoeuvre or a government's domestic stability. That does not require militarisation. A strategic fishery may never attract armed confrontation. It may still shape diplomatic posture, trade retaliation, licensing policy, and the tenor of wider relations with neighbours. Strategic value can reside in a cold store, a processing contract, or a seasonal employment cycle. It can sit in a ministry memo about regional unrest if plant closures continue. It can surface in a parliament where members from distant coastal districts insist that any quota compromise would amount to surrender. The frontier is political long before it becomes martial.

The Barents Sea shows the other side of the equation: how fish can matter strategically because they anchor both export value and regional employment, yet remain governed through durable co-operation rather than permanent confrontation. Norway and Russia have managed shared stocks there through the Joint Norwegian-Russian Fisheries Commission since 1976. That record cannot be understood by looking only at conservation. The Barents fisheries matter because they support fleets, processing, port activity, and coastal communities on both sides of the border. They also matter because failure in management would carry political costs well beyond the fishing industry. A trawler landing in Kirkenes or Murmansk is not merely unloading catch. It is feeding a chain of contracts, shifts, municipal revenues, and local expectations. The crate on the quay returns here in a different form: not as a symbol of dispute alone, but as a reminder that continuity itself has strategic weight.

Still, the Barents example should not be over-praised. Co-operation there does not erase conflict between economic priorities. It contains them. Large-scale export interests may favour efficiency, stable total catches, and the sort of predictability that supports investment in vessels and plants.

Small coastal communities may care more about where fish are landed, who owns quota, and whether the benefits of a resource remain local. Governments often speak as though these are complementary goals. Sometimes they are. Sometimes they are not. A quota system can produce impressive export performance while leaving local crews squeezed by leasing costs or by consolidation that shifts ownership away from the places where fishing culture remains strongest. "National success" can mask internal dispossession.

That is where food security returns with a different force. In much of the Arctic and sub-Arctic, especially in smaller and remote communities, the issue is not merely aggregate calorie supply. It is access, affordability, and continuity of local provisioning. Industrial fisheries policy may look sound at the national scale while worsening vulnerability at the community scale. A government can celebrate export strength while households face higher local prices, less dependable access, or the gradual erosion of small-scale fisheries that have social and cultural value beyond market metrics. The distinction between fish as commodity and fish as food is therefore not sentimental. It is administrative. It changes which harms are seen and which are buried inside national totals.

A sensible objection appears here. Many Arctic fisheries are deeply embedded in global trade, and some of the highest-value species are not staples for domestic consumption. In such cases, talk of food security can become inflated or opportunistic, a rhetorical cloak for commercial claims. That limitation is real. Not every insistence on "national food needs" deserves equal credence, particularly where the species in question is overwhelmingly exported or processed for distant markets. But the abuse of food-security language does not make the underlying issue trivial. It means analysts must ask a more precise question: whose food security, which species, what market channel, and what actual decision is at stake? Precision cuts through posturing.

Once these distinctions are made, negotiating behaviour becomes easier to read. A state that prizes domestic affordability may favour stable access and cautious management even when it sacrifices some short-term revenue. A state whose fishing sector depends heavily on export margins may tolerate more bargaining risk in order to defend a larger quota share or market access. A government under pressure from coastal labour may resist conservation measures less because it doubts the science than because it fears immediate unemployment in politically exposed regions. "Tolerance for risk" is not a mood. It is visible in choices: whether to accept an interim arrangement, whether to threaten unilateral action, whether to back port restrictions, whether to subsidise idled vessels, whether to keep a processing plant open through a poor season for political reasons rather than commercial ones.

This is why value-chain position matters so much. Catching fish is only one part of the struggle. States and firms that command processing capacity, cold storage, shipping links, and trusted market channels hold a quieter sort of power. They can absorb shocks better. They can shape where fish must go to become valuable. They can resist pressure longer. A fleet that lacks reliable landing options is weaker than a fleet with multiple ports and processors competing for its catch. A coastal state that fears closure of a single major plant will negotiate differently from one with a diversified industry. Strategic value accumulates at these bottlenecks. The politics of fisheries often turns less on what is in the water than on who can convert it into wages and tax receipts once it comes ashore.

And yet none of this should tempt us into treating fish as just another strategic commodity. They remain biological resources with rhythms, vulnerabilities, and distributions that resist neat political design. Their mobility frustrates the simple conversion of economic interest into stable control. A government may defend jobs in one season only to discover that the stock's path has shifted by the next. A processor built for certainty may inherit volatility instead. A quota defended as a matter of sovereignty may lose legitimacy if the fish become conspicuously absent from the waters used to justify it. Strategy in fisheries is therefore always under ecological correction. That is one reason it so often looks brittle.

The crucial point is modest but demanding. Once food, employment, exports, and regional politics are disentangled, fisheries disputes stop looking like technical quarrels with occasional patriotic decoration. They look like contests among different forms of dependence. One community depends on fish as dinner. Another depends on fish as wages. A state depends on fish as foreign earnings. A ministry depends on fish as proof that it can still defend inherited rights in a moving environment. Those dependencies do not automatically produce conflict, but they give conflict its structure. They tell us why one actor backs down and another does not, why one compromise looks sensible in a policy brief and impossible in a port town.

The question that follows is awkward and unavoidable. When a stock shifts and the old bargain weakens, which claim should weigh most heavily: the household worried about supply, the town worried about jobs, the exporter worried about access, or the government worried that conceding on fish today will narrow its room to govern tomorrow? The answer will not be found in the catch alone.

The book's framework and method

In October 2018, in Ilulissat, Greenland, ten parties signed the Agreement to Prevent Unregulated High Seas Fisheries in the Central Arctic Ocean. The immediate outcome was plain: they chose restraint before a commercial fishery had fully formed. That decision is easy to praise in general terms. It is harder and more useful to ask what made such restraint possible there when similar prudence proves elusive elsewhere. The answer does not lie in rhetoric about shared concern alone. It lies in a more exacting set of conditions: who could do what, who stood to gain from delay or haste, and how much could actually be seen and proved in a remote sea. That triad matters. It sorts gesture from governance.

A fisheries dispute often arrives in public as a single event. A vessel is detained. A quota is announced unilaterally. A stock assessment is rejected. A minister speaks in the language of rights. Headlines prefer the flashpoint because flashpoints are legible. But incidents mislead when treated as causes rather than symptoms. They tempt the observer to ask who escalated first, who broke ranks, who looked toughest. Better questions usually sit beneath the surface. Could the relevant authority monitor the activity it claimed to regulate? Did the parties face incentives that rewarded compromise or rewarded brinkmanship? Did the available evidence permit attribution strong enough to make the sanction credible? The wake tells a story.

That is the organising frame used here: governance capacity, incentives, and surveillance limits. The terms are simple enough to travel across cases without pretending that all fisheries are alike. They do not erase ecological variation, legal difference, domestic politics, or the weight of history. They discipline attention. Governance capacity asks what institutions can actually execute, not merely what they are entitled to declare. Incentives ask what fleets, firms, regulators, and governments gain or lose from compliance, delay, concealment, or confrontation. Surveillance limits ask what behaviour can be detected, interpreted, documented, and connected to a sanction that will hold. Put together, these three dimensions make distant-water disputes less mysterious. They also make them less theatrical.

Governance capacity is often flattened into a question of legal authority. That is too thin. A coastal state may possess extensive rights within its zone and still struggle to translate them into effective control. Capacity includes patrol reach, yes, but also licensing design, inter-agency co-ordination, data handling, port inspection routines, prosecutorial follow-through, and the bureaucratic patience required to keep ordinary rules from collapsing under seasonal pressure. A ministry with clear statutory powers but weak case management may detect violations without converting them into penalties. A coast guard may board successfully but deliver poor evidentiary records. A fisheries agency may produce regulations that field officers cannot apply consistently. Capacity, in other words, is visible in decisions and workflows. It appears in whether a vessel can be checked in time, whether its declarations can be compared against independent records, whether a sanction survives appeal, and whether different offices are working from the same factual picture. The law on paper matters. Administrative traction matters more.

The mackerel dispute in the northeast Atlantic after 2010 demonstrates the point. Iceland and the Faroe Islands argued that changing distribution justified larger shares. The European Union and Norway resisted what they saw as destabilising unilateral claims. Public argument revolved around fairness, science, and rights. Yet the dispute endured because no party could simply impose a stable settlement on the others. Regional institutions remained available, but consensus thinned. Scientific advice could inform total catches without determining allocation. Market measures, including later European restrictions against the Faroe Islands in 2013, exerted pressure, but they did not dissolve the underlying bargaining conflict. Seen through the present frame, the case was not merely a clash of narratives. It was a case of partial capacity. Institutions were strong enough to register disagreement and impose some costs, but not strong enough to convert ecological movement into an accepted revision formula.

Incentives explain why that kind of partial capacity so often produces friction rather than orderly compromise. Fisheries actors do not respond to rules in the abstract. They respond to the payoffs created by quotas, access windows, fuel costs, landing opportunities, processing contracts, domestic politics, and expectations about what others will do. A fisher who believes early capture may strengthen future allocation claims behaves differently from one who trusts that revisions will occur transparently. A government facing pressure from processors and coastal constituencies may prefer a risky negotiating stance to a prudent compromise if retreat looks like surrender at home. An operator who expects that paperwork anomalies will go unpunished has reasons to treat compliance as optional. Incentives are therefore not motives in the psychological sense. They are structured advantages attached to visible choices.

This is why the language of compliance needs care. Compliance does not mean moral agreement with the rule, nor does it mean perfect behaviour in all conditions. In fisheries governance, compliance is better understood as conduct sufficiently aligned with applicable rules that authorities can treat the system as functioning. The measure is practical. Are vessels transmitting as required? Are catches declared accurately enough for management to remain credible? Are landing records and observer accounts broadly consistent? Are restrictions obeyed without routine evasive testing? Once compliance is defined this way, the analytical task changes. One asks less whether actors respect the law in principle and more whether the institutional setting makes rule-following the easier and more rational path. That is where incentives and capacity meet.

Deterrence, too, is often overstated or misdescribed. In popular security language, deterrence can sound like a show of force that scares misconduct into retreat. Fisheries governance rarely works like that. Deterrence here is less theatrical and more cumulative. It exists when likely violations face enough chance of detection, enough probability of attribution, and enough cost at the end of the process that most operators prefer not to test the boundary too hard. A patrol vessel contributes to deterrence. So do port denials, documentation checks, licence suspensions, insurance requirements, and the reputational consequences of being marked as unreliable by buyers or regulators. Effective deterrence is therefore not a single threat. It is a believable sequence. Break the sequence at any point and behaviour changes.

The Joint Norwegian-Russian Fisheries Commission, established in 1976, offers a valuable counterweight to more dramatic cases. It should not be treated as a perfect model, still less as proof that shared stocks are easy to govern in northern waters. Its importance lies elsewhere. It shows how routine institutional practice can reduce the burden placed on spectacular enforcement. Joint scientific work, repeated negotiation, technical measures, and long familiarity between agencies do not abolish incentives to press advantage, but they alter them. They make defection more legible and, often, more expensive politically. They also widen the range of procedural responses before a disagreement becomes a public crisis. This is deterrence in a quieter key. Not a single act of intimidation, but a durable expectation that behaviour will be noticed, compared, and answered.

Still, surveillance limits complicate every element of this account. To govern fisheries is to govern partly through sight, and sight in distant seas is never complete. Vessel monitoring systems, automatic identification signals, satellite imagery, logbooks, observers, cameras, and port records all produce fragments. Some fragments align cleanly. Others do not. A gap in transmission may reflect equipment failure, deliberate interference, or a rule that did not require continuous broadcast in the first place. A transhipment record may be lawful yet still obscure catch origin. A suspicious route may indicate strategic behaviour or merely weather avoidance. Surveillance is therefore not the same as knowledge. It is an effort to narrow uncertainty enough to support action.

That distinction matters because many contemporary discussions slide too quickly from detection to certainty. They assume that once a vessel can be tracked, compliance can be guaranteed. But fisheries evidence is often inferential. Identity can be obscured by reflagging, complex ownership structures, call-sign confusion, or signal manipulation. At-sea conduct may remain partly hidden even when a vessel's location is known. Data systems may not speak to each other. Analysts may lack time or authority to turn anomalies into cases. A state can purchase expensive monitoring infrastructure and still discover that its weak point lies in integration, interpretation, or legal follow-through. Surveillance limits are therefore institutional limits as much as technological ones. What cannot be seen is only part of the problem. What cannot be proved, shared, or acted upon may matter more.

This is where the term "grey" activity becomes useful, provided it is used precisely. Grey activity does not mean everything suspicious. It refers to conduct that exploits ambiguity between clearly lawful and clearly unlawful behaviour. A vessel may fish aggressively near a boundary without crossing it. A company may rearrange ownership or flag registration in ways that complicate accountability without breaching an obvious prohibition. Operators may comply slowly, selectively, or legalistically, making enforcement costly while preserving plausible deniability. The value of greyness lies exactly there. It raises the evidentiary burden on authorities and the political burden on states considering escalation. Grey activity flourishes where incentives reward testing, surveillance remains partial, and formal capacity cannot easily convert suspicion into sanction.

A limitation should be admitted. This framework does not explain everything. Some disputes turn decisively on domestic elections, leadership style, historical grievance, or external geopolitical rupture that exceeds fisheries policy as such. Ecological change can also produce outcomes too abrupt or too poorly understood for institutional variables alone to clarify. A stock collapse, a sudden diplomatic freeze, or a wider sanctions regime may reorder behaviour in ways the triad only partly captures. That is not a reason to abandon the framework. It is a reason to use it with discipline. Capacity, incentives, and surveillance limits do not replace context. They prevent context from dissolving into anecdote.

That comparative discipline matters because Arctic and sub-Arctic fisheries invite false equivalence. A highly capitalised export fishery, a mixed coastal fishery, and a contested high-seas pocket should not be read as though they were versions of the same problem with different weather. Nor should singular incidents be mistaken for typologies. The method here compares not by pretending similarity where none exists, but by asking common questions across unlike settings. What are the operational limits of enforcement? Which actors gain from ambiguity? What evidence chain links behaviour to sanction? Which institutions provide routine off-ramps before conflict hardens? Such questions allow comparison without flattening difference. They also resist the lure of sensational cases that dominate attention precisely because the slow institutional failures behind them were ignored.

The preference for institutions and behaviours over dramatic moments is not an academic scruple. It is a practical necessity. Most fisheries conflict does not begin with extraordinary conduct. It begins with ordinary slippage: underpowered inspection, inconsistent reporting, delayed response, unresolved allocation formulas, weak follow-through after detection, domestic rhetoric that narrows room for compromise. These are not exciting features. They are the governing reality. They determine whether a later incident will remain containable or become a test of state resolve. The wake returns here as a quiet motif because institutions often leave their mark indirectly. One sees them in the steadiness of compliance, in the rarity of escalation, in the routine willingness of actors to submit disputes to procedure rather than to spectacle.

Trust belongs in this picture, but not as a soft virtue floating above hard politics. Trust in fisheries governance is visible conduct. It appears when parties share data without routine accusations of manipulation, when observer records are accepted as probative, when interim arrangements are honoured through a tense season, when inspectors believe that a counterpart agency will act on a referral rather than bury it. Distrust is equally visible. It produces duplicate checks, hedged commitments, unilateral announcements, and insistence on public signalling where private procedure once sufficed. Trust does not remove conflict. It lowers transaction costs and broadens the menu of tolerable compromise. Once damaged, it pushes actors back towards hard proof, strategic ambiguity, and the temptation to make examples of one another.

A final merit of this method lies in what it refuses. It refuses the comforting notion that better science alone resolves distributional disputes. It refuses the opposite temptation to read every fisheries disagreement as disguised power politics. It refuses to treat technology as salvation when institutions cannot digest the information technology produces. Above all, it refuses the glamour of crisis language when most of the real work of order lies in routine monitoring, procedural credibility, and the patient alignment of incentives with rules. Such refusals narrow the drama, but they sharpen the diagnosis.

The pressing test, then, is not whether a state can declare control over a moving fishery. Many can. It is whether control can survive contact with ordinary behaviour: a contested quota, a boundary-hugging vessel, a weak evidence chain, a season compressed by weather, an operator willing to live in the grey, a ministry reluctant to concede, a monitoring system rich in data and poor in closure. When these pressures gather, what deserves more weight - the headline incident that demands immediate reaction, or the quieter institutional pattern that explains why the incident became possible at all?

Chapter 2

Warming Seas and Stock Migration

A trawler leaves harbour under rules written for a familiar season. The crew know roughly when the fish should appear, where the water should cool, and when a trip becomes worth the fuel and risk. Then the pattern slips. The target stock arrives later than expected in one place, earlier in another, thinner on one ground and suddenly dense on a route that used to be marginal. The logbook still has boxes to fill. The quota still has dates attached to it. The inspectors still have patrol plans. But the fish are keeping a different calendar.

That is the governing disturbance. Not change alone, but change that arrives unevenly enough to unsettle habits before it produces a new stability. A stock does not simply march north in a straight line, neatly announcing a fresh political order. It stretches, contracts, lingers, doubles back, overlaps with other species, and alters the sequence in which different fleets encounter it. A warm patch here, a prey shift there, a salinity change that matters more than anyone expected - these are enough to turn a settled fishery into a moving target for law, science, and diplomacy. The sea does not move as one thing. Neither do the fish.

The central tension lies here: management systems need categories, but fish respond to conditions. States define zones, seasons, and allocation formulas because administration requires edges and intervals. Science also depends on comparability over time. Yet the biological reality that those systems are meant to govern can become less orderly precisely when political stakes rise. Once that happens, every practical question acquires a distributive shadow. A survey result becomes an argument about shares. A seasonal shift becomes a dispute about access. A bycatch rule becomes a disguised contest over who gets to keep fishing when the species mix changes faster than the regulation can follow.

This matters because warming seas do not merely alter abundance. They alter sequence. Who meets the fish first, who can catch them most efficiently, who lands them into domestic plants, who claims a new share, who insists that an old baseline still counts - all this can change even when no one is facing immediate biological collapse. That is why the politics of stock migration tends to feel so raw. The parties are not only debating conservation. They are deciding whether investments made under one geography should still command loyalty under another. A harbour built for a dependable run of fish, a fleet financed against familiar grounds, a quota justified by historic access: each becomes a stranded fact or a defended entitlement depending on where the stock now travels.

The pressure deepens when scientific advice loses its old certainty. Not because science becomes irrelevant, but because it is asked to do more than it can honestly do. Assessment models work best when the world they measure changes within known bounds. When productivity shifts, when survey coverage lags, when seasonal presence no longer mirrors earlier patterns, uncertainty stops being background noise and becomes politically charged. It enters negotiations in two forms at once. One side hears caution and thinks restraint. Another hears caution and suspects a tactic designed to freeze an older allocation in place. The clipboard matters here. So does who holds it.

And yet stock movement does not automatically produce breakdown. Some fisheries absorb shifting distributions with more poise than alarmist language suggests, especially where institutions can revise expectations without making every adjustment look like surrender. That limitation deserves respect. Not every altered migration pattern becomes a diplomatic quarrel; not every disagreement over surveys masks bad faith. Ecological systems have always varied, and some management regimes retain enough flexibility to accommodate movement without public crisis. But that is precisely the point. The difference lies not in whether the sea changes, but in whether the governing structure can treat movement as normal without inviting opportunism or resentment.

The stakes therefore reach beyond catch levels. They touch legitimacy. A rule remains credible only so long as those bound by it can still recognise the world it describes. If fish are turning up in new places, mixed catches are rising, and short seasonal windows are rewarding speed over paperwork, then a management regime designed for steadier distributions begins to look partial even when it remains lawful. Fishers notice first in route choices and deck decisions. Regulators notice in strained inspection schedules and rising arguments over declarations. Governments notice when technical disputes start attracting the language of injustice. Trust thins quickly under those conditions, because every uncertainty can be read as either a genuine problem or a strategic manoeuvre.

The path ahead follows that widening strain. It begins with the mechanisms that move stocks, not as a clean drift but as a set of uneven ecological signals that scramble timing and location. From there the argument turns to the unequal consequences of movement, where gain for one party often feels like dispossession for another and where disagreement over baselines hardens the bargaining climate. The pressure then passes through science under moving conditions, through the messy politics of mixed catches and bycatch, and into the compressed tempo of pulse fisheries where the real test of governance is whether rules can survive the busiest days rather than flatter the average season.

What deserves closest scrutiny, then, is not merely where the fish have gone. The sharper question is what kind of governing system can keep its authority when distribution, timing, and species overlap no longer fit the calendar, the survey grid, or the inherited bargain - and what happens when those mismatches arrive faster than institutions can admit.

Mechanisms of distribution change

In 2010, Iceland set a unilateral mackerel quota after years in which the stock had become far more available in Icelandic waters during the summer feeding season. The immediate political outcome was a bitter dispute with other coastal states, but the deeper lesson lay beneath the argument. The fish had not simply "moved north". They had altered the timing and geography of who encountered them, when, and in what density. A management quarrel that looked, in public, like a fight over shares was also a quarrel over mechanism. Why were the fish there? Would they remain? And which part of that shifting pattern should count when rights were discussed?

Those questions matter because range change in northern seas is often described too neatly. The popular image is a one-directional drift, as though warming water sends stocks marching poleward in orderly formation. Real fish do not oblige. They respond to temperature, certainly, but also to salinity, oxygen, prey availability, spawning grounds, depth, current structure, predator pressure, and the habits of different life stages. The frontier moves in patches. That is the point. What appears on a map as expansion may, at vessel level, feel like a broken sequence of arrivals, absences, detours, and brief concentrations that rewrite the season without announcing a new equilibrium.

Temperature remains the most familiar driver because it changes metabolic cost, habitat suitability, and the distribution of plankton and forage species on which larger fish depend. But temperature acts less like a command than like a filter. It does not tell a stock exactly where to go. It changes which waters remain energetically tolerable, which migration routes become efficient, and which feeding grounds reward the journey. For cold-water and sub-Arctic species, a shift of only part of the seasonal range can matter politically if it changes where commercially valuable fish become accessible. A stock does not need to abandon its old territory for governance to become unstable. It only needs to spend more time in a new place, or arrive there in a density that transforms what local fleets can catch before others do.

That is why salinity deserves more attention than it usually gets outside specialist circles. Salinity shapes water mass structure, egg and larval survival, prey assemblages, and the pathways by which fish move between feeding and spawning areas. In some northern seas, a change in salinity may alter the location of suitable nursery grounds or the depth at which prey becomes available, with consequences that are visible to fishers long before they are politically interpreted with confidence. A stock may appear to have "moved north" when the more exact description would be that different age classes are now occupying a different combination of depth, season, and feeding habitat because the water column itself has changed. Governance struggles because law likes surface boundaries while ecology keeps reorganising the vertical as well as the horizontal.

Prey dynamics make the picture messier still. Fish do not merely track preferred temperature; they track food under temperature constraints. If zooplankton blooms shift in timing, or if forage fish redistribute with changing conditions, predators may alter routes and seasonal residence even where the broader thermal envelope remains familiar. This is one reason migration often looks uneven rather than linear. A stock may extend into a new zone in summer feeding months yet still depend on old spawning grounds. It may thicken along one corridor but remain sparse elsewhere. It may seem to expand for several years, then partially retract when prey patterns change again. The practical implication is sobering. Management systems that interpret a temporary or seasonal concentration as permanent redistribution may over-correct, while systems that dismiss repeated new presence as mere noise may lag until resentment hardens.

The northeast Atlantic mackerel dispute is instructive precisely because it joined ecological complexity to political impatience. By the late 2000s and into 2010, Iceland and the Faroe Islands argued that the stock's summer presence in their waters justified larger quota shares. That position did not rest on fantasy. The fish were there, and in commercially meaningful quantities. Yet the argument over allocation immediately outran the biological observation. Was this a durable redistribution or a contingent feeding pattern? Should summer presence count as heavily as long-established spawning geography? How should repeated seasonal availability be weighed against historical catch? The science could illuminate the stock's behaviour, but it could not dictate the political value assigned to one part of the annual cycle over another. A warm season became a constitutional problem for the fishery.

The same logic applies to demersal stocks, though the mechanisms can be less visible to outsiders because bottom-associated species do not always advertise movement in the dramatic way pelagic disputes do. In the Barents Sea, cod distribution has long responded to changing environmental conditions, prey fields, and the broader structure of northern marine productivity. Fishers notice this not as an abstract range polygon but as altered catchability: where the fish can be found, at what depth, in what season, and with what fuel and gear costs. A stock may remain within a broadly familiar region while still reshaping the economics of access by shifting its centre of feeding or by changing the timing with which dense, commercially attractive concentrations become available to different fleets. The governing significance lies in encounter order. Catch it earlier and the whole season changes.

This is where "range expansion" becomes a dangerous simplification. The phrase sounds descriptive and harmless. In practice it can smuggle in two errors. It suggests, first, that a stock has moved in one consistent direction, and second, that movement can be read from occupancy alone. Neither is reliable. Fish may occupy a wider area without redistributing their most valuable biomass in a stable way. They may expand seasonally while retaining critical dependence on historical spawning grounds or migration corridors. They may shift northward in one life stage and remain stubbornly conservative in another. For governance, these nuances are not academic refinements. They determine whether a state is seeing a durable change in entitlement-relevant access or a repeated but still unstable pulse that should be handled with caution.

A good example of this tension appears whenever survey evidence begins to strain under moving baselines. Scientific assessment works by comparing the present to a meaningful past, but the past can become a poor guide when environmental conditions alter both fish behaviour and the tools used to observe it. Survey routes, timing, and catchability assumptions are usually designed around earlier distributions. When stocks spread into new areas, deeper waters, or different seasonal windows, the monitoring system itself may need revision. That revision takes time. Meanwhile, political actors treat uncertainty selectively. One side may present newly observed presence as decisive proof of redistribution. Another may argue that incomplete survey adaptation makes any revision premature. The clipboard returns here, but so does the calendar. Scientific disagreement often intensifies not because researchers are careless, but because the sea has made older routines less informative.

And yet the picture should not be overdrawn. Not every apparent environmental shift produces major redistribution, and not every stock responds quickly even when conditions change around it. Site fidelity, spawning behaviour, habitat dependence, predator interactions, and simple ecological inertia can slow or complicate movement. Some stocks remain more stable than popular narratives imply. Others fluctuate without establishing a clear new pattern. That limitation matters because it guards against a lazy determinism in which every warm year is treated as a jurisdictional revolution. Governance would become impossible if every anomaly triggered immediate renegotiation. The serious challenge is to distinguish between noise, oscillation, and meaningful redistribution without pretending that the distinction will always be clean in the moment.

The timing problem makes this harder. Stocks do not only change where they are; they change when they are there. Seasonal arrival can advance, delay, shorten, or stretch. A fishery planned around an old calendar may therefore miss the decisive window even if the stock remains broadly within the same legal space. This matters for quota administration because annual totals disguise temporal concentration. A state may technically have access to a stock over the course of a year while losing the commercially valuable part of the season to a neighbour that encounters dense schools earlier. It matters just as much for enforcement. Inspection schedules, patrol deployments, observer coverage, and port staffing are often arranged around average expectations. Fish do not care about administrative averages. If the season compresses into a shorter and more intense pulse, compliance pressure rises exactly when monitoring systems are most likely to be thin or mistimed.

Mixed-stock encounters deepen the strain. As distributions change, fisheries that once targeted one stock in relative separation may begin to overlap more heavily with another species that is protected, capped, or politically sensitive. This is not merely a biological inconvenience. It changes the meaning of bycatch rules. A fleet may insist it is pursuing a legitimate target while regulators suspect that mixed catches are being used to secure access to fish that would otherwise be restricted. Monitoring demands rise accordingly. More scrutiny falls on gear choice, haul composition, spatial closure compliance, and the credibility of logbooks. What began as a shift in prey or water conditions can end as a dispute over intent. Governance has to decide whether the overlap is an unfortunate ecological fact or an exploited ambiguity.

The underlying lesson is easy to state and hard to govern. Fish distributions change through interacting mechanisms, not through a single climatic command. Temperature may open opportunity, salinity may reshape habitat, prey may redraw the reward structure, and seasonality may determine whether any of that becomes politically explosive. These mechanisms do not simply move biomass around a chart. They rearrange access in time, by life stage, by depth, and by commercial usefulness. Once that happens, old bargains do not merely look outdated. They begin to look partial, because they continue to assign rights according to a geography that fish are obeying less faithfully than before.

The quiet motif throughout is the season itself. Not the year in the abstract, but the lived sequence of when fish arrive, when surveys occur, when fleets commit, and when regulators are ready. A stock may seem stable on an annual map while becoming politically disruptive on a seasonal one. That difference is exactly where governance is tested. It must decide what counts as a meaningful shift, which part of the stock's annual behaviour deserves regulatory weight, and how much instability institutions can absorb before delay starts rewarding unilateral action.

That decision cannot be postponed by repeating the language of northward movement as though it were explanation enough. When the next stock appears earlier in one zone, later in another, and in a species mix that unsettles both quota arithmetic and inspection planning, what should count as the real signal: expanded presence, changed timing, altered catchability, or the simple fact that the old calendar no longer describes the fishery with enough honesty to govern it?

Winners, losers, and new entrants

In August 2013, the European Union imposed trade measures against the Faroe Islands during the dispute over Atlanto-Scandian herring and north-east Atlantic mackerel. The immediate outcome was clear enough: Faroese exports of the species in question faced restrictions, and access to certain EU ports was curtailed. The quarrel looked, from a distance, like a technical disagreement over quota discipline. It was not. It was a distributional fight made sharper by ecological change. Fish had become more available in Faroese waters, and with that shift came a claim that the old allocation no longer matched the sea. A new share demanded recognition; an old share demanded defence. The harbour remembered everything.

That last point matters because fisheries disputes are never argued on a blank slate. Ports, freezing plants, repair yards, fuel depots, vessel loans, crew traditions, and local tax bases are built around expectations formed under earlier distributions. When stocks move, the biological event is immediate but the economic landscape remains stubbornly fixed. A fish that spends a longer feeding season in new waters may create new opportunities for one coastal state while leaving another with infrastructure designed for a geography that no longer delivers the same certainty. Access shifts faster than steel and concrete. That mismatch is where resentment starts to harden into politics.

The governing difficulty is easy to name and hard to resolve. Ecological change creates winners, losers, and aspirants at the same time. A state that previously stood at the edge of a fishery may now encounter the stock more often and in commercially useful density. Another may still hold historic rights, institutional seniority, and processing capacity, yet feel the practical grounds of its advantage slipping. A third may see an opening and argue that if the stock now traverses waters or seasons previously outside the effective bargain, then participation should widen. None of these positions is inherently absurd. Each can be defended by reference to some combination of fairness, investment, science, and need. The conflict begins when those justifications rest on different clocks.

Baseline is the crucial word. It sounds dry. It is not. A baseline decides which past counts. Is entitlement measured against historical catches established over decades, reflecting long-term participation and capital investment? Is it measured against current zonal presence, on the logic that fish found in one zone should generate corresponding access for that coastal state? Is it measured against a more recent period during which the stock has demonstrably changed its seasonal range? Each answer produces a different map of justice. Choose history, and established participants appear prudent and deserving. Choose present distribution, and earlier beneficiaries can look entrenched. Choose a recent transitional period, and everyone suspects that the other side is selecting dates to freeze temporary advantage into law.

The mackerel dispute involving Iceland after 2010 remains the most revealing example because it exposed all of these tensions at once. Iceland argued that the stock's growing summer presence in Icelandic waters made the established allocation indefensible. The claim did not depend on rhetoric alone; the fish were physically present in commercially significant concentrations. Yet the opposition to Iceland's stance, particularly from the European Union and Norway, also rested on more than simple refusal. Historic participation, prior agreements, and long-standing investment in fleets and processors all stood behind the older arrangement. The fight therefore turned on a deeper question: when the fish move, what exactly must move with them - biomass, quota, or only a promise to discuss revision later? The answer was never merely scientific. It was distributive.

A stock entering new waters does not only create a winner on paper. It changes who can catch first, who can justify new vessel activity, who can claim local economic benefit, and who can speak in the name of fairness with fresh political force. That is why new entrants often sound more morally confident than established participants find tolerable. The argument is intuitive: if the fish are here, why should an old bargain continue to exclude us or minimise our share? But established participants can produce an equally forceful reply. They financed vessels, built landing networks, hired labour, and accepted management disciplines under one set of assumptions. To revise shares too quickly may turn long investment into stranded cost. One side invokes present reality. The other invokes legitimate reliance. Both are describing a form of injury.

The harbour keeps score.

That quiet motif is worth holding because fisheries politics becomes sharper when ecological movement collides with immobile assets. A quota can, in principle, be renegotiated. A processing plant cannot pick itself up and follow a stock. A community built around generations of landings cannot easily be compensated by being told that the fish are now more abundant elsewhere. The losers from redistribution are therefore not only states in the abstract. They are places. Their losses become visible in idle shifts, shorter seasons, tighter margins, and regional political anger. Once those costs are felt ashore, compromise narrows. Governments begin to bargain not only with neighbours but with their own coastal constituencies, who rarely experience "adaptive governance" as anything other than diminished certainty.

This is one reason perceived unfairness grows faster than outsiders expect. The dispute is not simply over whether ecological change has occurred. It is over what earlier obedience should now be worth. States that accepted quotas, invested in monitoring, and structured industries around established arrangements often believe they are being punished for having behaved responsibly under the old rules. States gaining access often believe they are being asked to subsidise history by accepting a share that ignores present ecological fact. Neither side thinks of itself as opportunistic. Each sees the other as trying to socialise its own adjustment costs. Bargaining becomes emotionally charged because both parties can plausibly frame themselves as the aggrieved adult in the room.

The herring and mackerel disputes involving the Faroe Islands made this painfully visible. Faroese authorities argued that the stocks' changed distribution demanded a larger allocation. Their counterparts treated unilateral catches as destabilising and contrary to co-operative discipline. The eventual recourse to trade measures in 2013 showed how far such arguments can travel once the parties no longer trust that existing institutions will correct the distributional imbalance quickly enough. What started as a question of fish presence became a test of whether newcomers or newly advantaged participants could force revision through action rather than await consensus. Once market access enters the argument, the vocabulary of fairness acquires sharper edges. A contested stock becomes a contested livelihood.

The same logic appears in demersal fisheries, even where the politics are less publicly theatrical. When cod or another bottom-associated stock shifts its commercially useful range, the immediate consequence may be a changed order of access rather than an obvious crossing into entirely new legal space. One fleet now encounters dense concentrations earlier. Another must steam farther or wait longer. A port that once served as a dependable landing point receives less throughput, while a smaller or previously peripheral port becomes newly valuable. These are not minor adjustments. They alter bargaining power inside states as well as between them. Regions that gain new landings demand recognition. Regions that lose them demand protection. What looks internationally like a dispute over quota formula may domestically be a fight over which town still has a future.

A useful counter-example should be admitted. Not every shifting stock produces bitter distributive conflict, and not every new entrant is treated as a threat. Some management systems absorb redistribution more effectively because the relevant parties already possess routines for adjustment, data sharing, and linked concessions. The Joint Norwegian-Russian Fisheries Commission, established in 1976, has often managed shared northern stocks under conditions that could easily have produced worse distributive quarrels than they did. That record does not erase conflict or make the Barents a paradise of disinterested co-operation. It does show that institutional habit can soften the transition from ecological change to political renegotiation. Where parties trust that revision is possible without immediate humiliation, the arrival of new access does not have to provoke unilateralism.

Still, that limitation should not be overstated. Even comparatively durable institutions depend on an underlying willingness to treat shifting distributions as politically discussable rather than as a threat to settled identity. Once domestic politics sacralises a quota share or a species, room for accommodation shrinks sharply. A baseline ceases to be a technical reference point and becomes a moral claim about who has earned the right to count. Historical catch is then defended not merely because it reflects past exploitation, but because it symbolises prior discipline, sunk cost, and national standing. Zonal presence is claimed not merely because the fish are visibly present, but because denying that presence feels like a refusal to acknowledge reality itself. The argument moves from arithmetic to recognition.

Recognition is precisely what new entrants seek, whether they are genuinely new participants or long-marginal actors newly empowered by ecological change. They want the management system to admit that the fishery they are now experiencing is not the fishery on which the old deal rested. Established participants often hear this as a request for redistribution without corresponding responsibility, particularly if they fear that newcomers will claim a share without bearing equal conservation burdens or market discipline. The suspicion is predictable. So is the resentment on the other side, where waiting patiently for institutional revision can look like a request to subsidise the old order indefinitely. When both perceptions take hold, bargaining becomes slower, harsher, and more theatrical.

This is why ecological change so often spills into distributive conflict even when all parties publicly endorse sustainability. Sustainability does not tell governments how to divide gain and loss. It does not answer whether a processing region built under former abundance deserves protection against a stock's departure, or whether a newly favoured coastal state must remain bound by a bargain shaped before the fish became truly available to it. Those are political questions in the deepest sense: they concern the distribution of costs, opportunities, and recognition under altered material conditions. The biology triggers the dispute. Distribution gives it staying power.

The trouble deepens when parties disagree not only on principle but on evidence. A state that sees repeated seasonal presence in its waters may treat the matter as settled fact. Another may insist that the pattern remains unstable, partial, or too recent to justify major reallocation. Both can point to catches, survey results, and vessel behaviour. Both can accuse the other of selective memory. Baseline disputes are so intractable because they combine data with judgement. The data may show movement. Judgement decides which movement counts, over which period, for which purpose. Once quota size and quota share are both in play, uncertainty becomes politically expensive. Nobody wants to concede present catch for fear that it becomes tomorrow's precedent.

A further complication lies in the difference between access and benefit. A stock may become newly available to a coastal state, yet that state may still lack the vessels, quota architecture, processing base, or market routes needed to convert ecological gain into durable advantage. The reverse can also be true: an established participant may retain enough capital and infrastructure to profit from a stock even as its local ecological access worsens. This means winners and losers are not always identical with gainers and decliners in the water. Some actors win because they control the chain that turns fish into money. Others lose because they cannot scale quickly enough, even when the biology seems to favour them. Distributional conflict therefore extends beyond the moment of capture. It runs through ports, factories, contracts, and the state's own capacity to translate changing abundance into recognised rights.

That broader view makes the politics easier to read. Bargaining is hardest where ecological change is obvious enough to provoke demands, but not settled enough to settle them; where investments made under older distributions remain costly and politically visible; and where the baseline itself has become a disguised fight over whose history deserves legal respect. The injustice each side sees is real, even when the remedies proposed are self-serving. One side sees exclusion from fish it can now visibly catch. The other sees the devaluation of institutions, capital, and restraint built under earlier conditions. The management problem becomes a constitutional one for the fishery: how to revise without making obedience to the old order look foolish, and how to preserve continuity without making the new order look fraudulent.

The next dispute over a moving stock will almost certainly be presented as a disagreement over science, sustainability, or procedure. Part of it will be exactly that. But the sharper test sits underneath. When fish create new winners and expose new losers, which past should govern the bargain: the one remembered in the harbour, the one measured in the latest season, or the one a state hopes to secure before the next movement changes the argument again?

Scientific advice under moving baselines

In 1902, representatives of northern European states met in Copenhagen to establish what became the International Council for the Exploration of the Sea. The outcome was both practical and ambitious: they created a standing scientific body for seas already understood to be shared, contested, and economically vital. More than a century later, that founding moment still casts a long shadow over fisheries politics. It reminds us that science entered marine governance not as an ornament, but as an attempt to make common judgment possible where no single state could see the whole fishery alone. Yet the sea that science was asked to measure has become less obedient to the habits by which it was once measured.

A stock assessment begins with a simple hope. Measure carefully enough and the argument will narrow. Count fish, compare catches, estimate recruitment, track spawning biomass, adjust for mortality, and a reasonable total allowable catch should follow. That hope remains indispensable. It is also incomplete. When stocks shift their timing, stretch into new survey areas, alter productivity, or become differently catchable under changing conditions, scientific advice does not stop mattering. It becomes more burdened. The survey line bends.

That burden arises because fisheries science depends on comparability over time. Models are not crystal balls; they are disciplined ways of relating present evidence to past patterns. Survey design, sampling assumptions, maturity schedules, growth rates, and catchability estimates all rely on the idea that yesterday remains a useful guide to today. When warming seas disturb that relationship, the problem is not ignorance in the crude sense. It is a model strain. A stock may still be measurable, but the meaning of the measurement becomes less secure. If fish occupy new waters during surveys, if productivity changes under altered temperature and prey conditions, or if seasonal behaviour shifts so that old sampling windows no longer capture the relevant part of the stock, scientific advice begins to carry wider uncertainty even when researchers are doing careful work.

This is where "moving baselines" becomes more than a clever phrase. A baseline in fisheries science is not merely a remembered past. It is the operational reference against which abundance, productivity, and risk are judged. If the stock's distribution, life history, or relationship to the environment changes, the baseline may quietly cease to describe the system with enough honesty. The result is not simply worse science. It is contested science. One party argues that the model undercounts fish because survey coverage lags behind new distribution. Another argues that unusual recent abundance in one area reflects temporary environmental conditions rather than durable redistribution. A third insists that precaution demands restraint precisely because the science is under stress. Each claim may contain truth. None escapes politics once quota size and quota share are both on the table.

The north-east Atlantic mackerel dispute exposed this tension with unusual clarity. As the stock became more available in Icelandic and Faroese waters in the late 2000s and early 2010s, scientific advice remained central to public argument. But the meaning attached to that advice diverged sharply. Coastal states could broadly acknowledge that the stock's distribution had changed while still disagreeing over how much significance to assign to that change for allocation. Scientific work could inform overall catch advice; it could not settle the prior quarrel over which part of the stock's behaviour deserved regulatory weight. Summer feeding presence, historical catches, spawning geography, and long-term participation all competed as candidate baselines. "More science" did not dissolve the conflict because the conflict was not only about abundance. It was also about entitlement under altered ecological conditions.

That distinction matters. Fisheries science is often asked to answer two questions at once: how much fishing is biologically prudent, and how the resulting opportunity should be divided. The first is difficult enough. The second exceeds science's authority, however rigorous the data. Yet the two questions become entangled in practice because uncertainty about stock size directly affects the distributional argument. If the recommended catch is lower, allocation becomes harsher. If the stock appears more productive than previously assumed, pressure grows to revise shares in line with new access. Scientific uncertainty therefore acquires political salience not merely because it clouds conservation, but because it magnifies the consequences of every allocation formula already under strain.

Consider the Barents Sea, where the Joint Norwegian-Russian Fisheries Commission has long provided a comparatively durable setting for shared-stock management. The relative stability of that arrangement does not mean the science has been easy. Northern ecosystems remain variable, survey conditions remain demanding, and stock productivity can shift. What matters is that repeated co-operation has built enough procedural trust for disagreement over evidence to remain discussable rather than instantly destabilising. Joint work does not eliminate argument. It changes the form of argument. Parties can contest assumptions, compare observations, and revise methods within an institutional relationship that is expected to endure. Scientific advice still strains under environmental change, but the strain is less likely to become immediate diplomatic theatre because the process itself carries legitimacy.

This example is useful precisely because it prevents a common error. One might be tempted to say that the answer to moving baselines is simply better data and closer scientific co-operation. Better data do help. Co-operation matters enormously. And yet neither is self-executing. Scientific institutions do not float above the bargaining field; they sit inside it. Survey design requires money, vessel time, access to waters, political permission, and agreements about what counts as acceptable evidence. A joint assessment gains authority only if the parties are willing to let its conclusions discipline their claims. When that willingness weakens, the scientific process can remain formally intact while losing some of its power to settle anything important. The survey line bends again, this time under institutional rather than ecological pressure.

Changing productivity poses an especially hard problem because it unsettles the hidden promise that management can infer future resilience from past response. A stock may produce strong recruitment under one set of environmental conditions and quite a different pattern under another. Growth rates may shift. Spatial overlap between life stages and survey effort may change. Predator-prey relations may alter mortality in ways that older model structures capture poorly. None of this means that stock assessment becomes futile. It does mean that managers and negotiators who treat historical productivity as a stable backdrop may be working with a false calm. Scientific advice under such conditions often contains caveats, sensitivity analyses, and warnings about interpretation. Those caveats are not weakness. They are honesty. Politically, however, honesty can look like vulnerability when different parties stand ready to use uncertainty selectively.

The danger becomes sharper when survey coverage and stock distribution no longer match. Fish do not rearrange themselves for the convenience of research cruises. A stock extending into new areas, occupying different depths, or altering seasonal presence may fall partly outside the established survey frame. That creates a familiar dispute. One side argues that the assessment underestimates biomass because the fish have moved beyond the old observation field. Another warns that recent concentrations in newly observed areas should not be mistaken for a wholesale redistribution of the stock. Both positions may be scientifically arguable, especially in transition periods. The governance consequence is severe. If parties no longer trust that the measurement system captures the fishery they are actually living with, every quota discussion becomes partly an argument over whether the ruler itself is straight.

A further complication lies in the difference between abundance and accessibility. Scientific surveys may detect fish; fleets care when fish become catchable at commercially meaningful times and places. A stock can remain broadly healthy while its availability to different states changes in ways the assessment does not fully translate into allocation politics. This creates a subtle but potent mismatch. Scientists may speak responsibly about stock-wide dynamics, while governments and industry actors experience highly localised gains or losses. One ministry sees vessels travelling farther for the same catch. Another sees new landings arriving at domestic ports. Both then approach the science with different appetites. Scientific advice becomes not only a technical resource but a political instrument through which each side tries to make its lived geography look like the authoritative one.

Bycatch and mixed-stock settings deepen the problem. Where a target stock increasingly overlaps with a capped or protected stock, assessment uncertainty no longer concerns only total abundance. It affects the operational credibility of the entire fishery. If regulators suspect that fishers are using uncertainty around species mix, spatial overlap, or discard patterns to stretch access, science is pulled into enforcement. More sampling may be demanded. More scrutiny falls on logbooks, observer data, and electronic monitoring. Disagreement over stock status can then spill into disagreement over conduct. A party may say the model is too conservative. Another may hear a defence of misreporting. Scientific uncertainty, once confined to advisory tables, starts shaping who is treated as trustworthy on deck and in port.

Still, a limitation should be recognised. Not every case of scientific disagreement under moving baselines reflects bad faith, and not every management system collapses when the ecology shifts. Some disputes look more dramatic than they are because public debate compresses complex methodological discussion into accusations of bias. Scientists can and do revise models, expand surveys, and improve understanding over time. Fisheries institutions are not helpless before change. The reason to dwell on uncertainty is not to discredit expertise. It is to show that expertise does not enter politics as a final word. It enters as disciplined evidence interpreted within institutions whose members have interests, memories, and constituencies.

That is why trust matters so much, and why it must be described carefully. Trust here is not sentiment. It is visible practice. It appears when parties accept joint survey results even when those results are inconvenient. It appears when methodological disputes are carried through working groups rather than instantly translated into ministerial accusation. It appears when a temporary shortfall in data does not trigger unilateral quota claims because the parties believe revision remains possible through procedure. Distrust is equally visible. It shows up in competing national assessments, in efforts to publicise only the most favourable interpretation of uncertainty, in procedural delay used as bargaining pressure, and in the hardening belief that the other side wants science only when science flatters its position.

The phrase "more science" often works politically because it sounds neutral. Who could object to improved knowledge? But the phrase hides a choice. More science of what kind, designed by whom, funded by whom, validated under which assumptions, and linked to what decision rule? A richer dataset will not produce agreement if the parties remain divided over which baseline deserves normative force. Nor will methodological refinement alone solve disputes when revised evidence threatens to redistribute value among states, fleets, or regions. Science can narrow uncertainty, expose weak claims, and discipline opportunism. It cannot decide how much historical investment ought to count against present ecological reality. That judgment belongs to institutions, and institutions cannot borrow legitimacy from science indefinitely if they refuse to confront distribution openly.

The larger lesson is not cynical. It is exacting. Scientific advice under moving baselines is most powerful when it is embedded in procedures that separate, as far as possible, the question of what the stock appears to be doing from the question of how parties will absorb the consequences. Where those questions collapse into each other, uncertainty becomes toxic. Every revision looks strategic. Every caveat looks partisan. Every survey gap looks like an opportunity for someone else. The ruler no longer measures only fish. It measures patience.

That is the test that follows. When productivity shifts, survey coverage lags, and stock behaviour ceases to fit the assumptions that once made assessment feel authoritative, what deserves more protection: the integrity of cautious science, the urgency of revising allocations to match new access, or the institutional trust required to keep disagreement from turning every scientific update into a diplomatic provocation?

Mixed-stock fisheries and bycatch politics

In December 2010, the North Pacific Fishery Management Council approved a hard cap on Chinook salmon bycatch in the Bering Sea pollock fishery. The outcome was immediate and concrete: a fishery targeting one of the world's great industrial whitefish resources would now operate under a rule shaped by another species altogether, one carrying ecological, cultural, and political weight far beyond its incidental presence in a trawl net. That decision revealed something larger than a technical adjustment. Bycatch was no longer a side issue. It had become the line through which access itself would be argued.

This is how mixed-stock politics usually hardens. A fleet sets out after a lawful target species. The water does not separate itself neatly for regulatory convenience. Fish overlap by season, depth, feeding grounds, and migration route. A haul that is commercially rational for one stock may carry another that is protected, weak, capped, or symbolically untouchable. The official language calls this bycatch. The political language quickly becomes harsher. Was the overlap unavoidable? Were operators steering carefully enough? Is the protected species genuinely incidental, or is the "target" fishery serving as a cover for tolerating higher mortality in a stock that regulators are under pressure to defend? One net, two arguments.

The basic problem is structural. Management units are administrative objects; fish assemblages are ecological ones. Regulators divide stocks into zones, seasons, quotas, and technical categories because institutions require lines. Fish respond instead to temperature, prey, currents, life stage, and habitat. When warming seas alter these relationships, overlap increases in places where the old categories once looked serviceable. A target stock enters new waters, but so does a sensitive companion species. A fishery previously understood as relatively clean becomes politically exposed because its catches now include more of the wrong fish, at the wrong time, under the wrong reporting assumptions. Bycatch then stops being a biological residue and becomes a constitutional problem for the fishery.

The Bering Sea pollock and Chinook salmon case remains useful because it shows the full transformation from incidental capture to proxy battle. Pollock fleets argued, often with reason, that mixed catches were shaped by complex and shifting ecological conditions. Salmon advocates, many subsistence users, and managers concerned about vulnerable runs treated the same catches as evidence that industrial efficiency could no longer be insulated from wider ecological responsibility. The resulting cap did more than constrain bycatch. It redistributed risk. A vessel targeting pollock now had to think about salmon encounters not as unfortunate noise but as a possible choke on continued access. The significance of the rule lay precisely there. A fishery for one stock could be curtailed because of another. Once that happens, the politics of bycatch becomes the politics of entry.

That phrase, choke stock, matters. It captures the moment when a fishery is not limited by its target quota but by the incidental capture of something else. In mixed-stock settings, the practical ceiling on effort may therefore be set by the species least available for lawful mortality. This changes behaviour. Operators may lobby for spatial flexibility, gear exemptions, or revised accounting. Regulators may impose closures, move-on rules, observer coverage, or electronic monitoring. Environmental groups may argue that the target fishery is effectively too blunt for the stock mixture now present. None of this turns only on conservation principle. It also turns on who gets to keep fishing once overlap makes innocence hard to prove.

A second example sits farther south in geography but close to the same governing logic: the North Sea mixed demersal fisheries under the European landing obligation introduced from 2019. Fleets targeting haddock, saithe, or other commercially valuable demersal species could find their effective freedom constrained by cod, a weaker stock whose bycatch risk threatened to close or sharply limit activity. The dispute was not only about discards or technical compliance. It was about whether a fishery should be governed by the composition of an actual haul or by the paperwork category of the vessel's declared target. Fishers argued that ecological reality made perfectly clean fishing impossible in a mixed fishery. Regulators and conservation advocates responded that this was precisely why stricter accountability was needed. Again the same conflict surfaced: was bycatch an unavoidable ecological fact, or a negotiated tolerance for continuing access?

That tension becomes sharper when distributions shift. A warming sea does not merely move the target stock. It can redraw the overlap zone between target and protected species, or alter the season in which that overlap becomes commercially consequential. A fleet that once knew when and where to avoid a sensitive stock may discover that the old timing no longer works. A spawning aggregation appears earlier. A forage concentration draws multiple species together in a new area. A demersal fishery operating under historical assumptions now encounters protected juveniles or capped bycatch at rates that make old measures look naive. Regulators, faced with that change, rarely treat it as neutral ecology. They read it through behaviour. Are vessels adapting? Are they shifting grounds voluntarily? Are catches being reported honestly? The moment suspicion enters, monitoring demands rise fast.

This is why bycatch rules so often become proxy battles over access. Officially, the argument concerns mortality on the sensitive stock. Operationally, everyone understands that the real prize is permission to continue fishing the target species. A party that cannot win a larger target quota may still constrain a rival fishery by insisting on tighter bycatch protections. A regulator uncertain about outright closures may use bycatch caps as a disciplined way to force effort reduction without relitigating the whole management regime. Fishers, for their part, may frame bycatch limits as disguised exclusion, particularly where they believe their vessels are being held to a standard no mixed fishery can meet consistently. The protected stock thus acquires a second life in politics. It becomes the argument through which access to another stock is widened, narrowed, or made conditional.

Trust thins quickly here because bycatch is hard to observe cleanly in real time. A quota can be counted at landing. A bycatch event often has to be inferred from what happened in a specific tow, in a specific place, under conditions that may not be directly visible to regulators. Were fish discarded before reporting? Was species identification correct? Did the crew sort accurately when the deck was crowded and the weather poor? Did the vessel remain in an area after a warning sign that should have triggered a move? These are not philosophical questions. They attach to visible behaviour: camera placement, observer notes, electronic logbook entries, move-on compliance, net configuration, haul timing. Once management depends on such details, the politics of evidence becomes as important as the biology of overlap.

The result is an asymmetry of proof. Fishers often know more about the immediate ecological complexity of a haul than any regulator ever will. Regulators, however, must build rules for fleets they cannot simply trust on declaration. When a protected or tightly capped stock becomes a choke point, this asymmetry becomes intolerable to both sides. Fishers resent being judged through coarse categories that fail to capture the real fluidity of the sea. Regulators resent being asked to take innocence on faith when incentives plainly favour understatement. Monitoring expands to fill the gap. More observers. More cameras. More audit trails. More requirements to document not just what was landed, but what was encountered, sorted, discarded, and avoided. Efficiency suffers, but not accidentally. The friction is the policy.

A sensible objection should be admitted. Mixed-stock fishing is not always a story of evasion, and rising bycatch does not prove bad faith. Ecological overlap can outpace regulatory categories without any deliberate wrongdoing by vessels. Fishers may make genuine avoidance efforts in conditions where stock mixtures change faster than route planning, especially in high-latitude fisheries already strained by distance and weather. Some bycatch rules, if drafted crudely, can punish honest operators while doing little to improve ecological outcomes. That limitation matters because it prevents a familiar collapse into moral simplification. Not every bycatch dispute pits villains against guardians. Sometimes it pits ecological reality against a management architecture that no longer fits it.

And yet that concession does not remove the political charge. The very ambiguity that makes mixed-stock fishing hard to govern is what makes it useful as a field of strategic argument. A fleet under pressure may insist that incidental mortality is unavoidable unless the entire fishery is closed, a conclusion regulators may view as exaggerated bargaining. A regulator may tighten rules in the name of precaution, while industry reads the move as a way of reallocating opportunity without openly admitting it. Communities dependent on the target fishery see livelihood under threat; communities attached to the protected stock see delayed justice finally becoming tangible. Both sides can speak sincerely. Both also understand that the bycatch rule will decide who bears the economic adjustment.

This is especially true where management units fail to match ecological reality. A stock may be assessed in one way, encountered in another, and regulated through a third. Juveniles of a protected species may overlap with an industrial fishery that was designed around adult distribution. A pelagic stock may pass through demersal grounds under conditions that make target declarations legally coherent but ecologically misleading. Seasonal closures may be calibrated to older migration patterns and therefore miss the new overlap window. Once such mismatches multiply, management starts chasing the fishery rather than governing it. Bycatch politics then becomes an argument about institutional lag. Who must absorb the cost of categories that no longer describe the sea?

A quiet motif runs through all of this: the net. It is both simple and uncompromising. A regulation may distinguish target from incidental species with legal precision, but the net does not read the rulebook. It gathers what the water and the gear bring together. Governance begins only after that fact. The serious question is how much of the burden should fall on fishers to avoid ecologically mixed water, and how much should fall on institutions to admit that their own units and seasons no longer correspond to what the net is actually finding. Pretending the categories still fit only raises the eventual cost of enforcement.

The move from trust to verification is therefore not a sign of policy maturity alone. It is also a symptom of institutional stress. A management regime that once relied on logbooks and periodic checks may shift towards near-continuous observation because contested bycatch makes ordinary reporting politically insufficient. This can improve accountability. It can also provoke its own backlash, especially where crews and operators feel they are being monitored not because rules are clear, but because regulators no longer know how else to govern a fishery whose ecological composition has slipped beyond old assumptions. Monitoring then ceases to feel like neutral administration. It feels like a verdict delivered in advance.

What makes mixed-stock fisheries so combustible is that they sit at the junction of biology, evidence, and entitlement. Overlap is ecological. Bycatch accounting is administrative. Access is political. A tightly capped stock can thus determine the fate of an entirely different fishery, not because anyone planned it that way, but because the sea has fused what institutions preferred to keep separate. When that happens, disputes over gear, observer coverage, move-on rules, and bycatch caps are never merely technical. They are the visible surface of a deeper argument about who should be allowed to keep fishing under changed ecological conditions.

The unresolved test is hard to escape. When a target fishery increasingly overlaps with a protected or tightly capped stock, what should count as fair governance: stricter monitoring and tighter bycatch limits that risk turning ordinary fishing into a near-forensic exercise, or looser tolerance that may preserve access in the short term while quietly normalising mortality no one is prepared to defend in public?

Timing, pulse fisheries, and enforcement strain

In January 2019, the Norwegian Coast Guard intercepted the Russian trawler *Melkart-5* in the protection zone around Svalbard, and the incident quickly sharpened into a legal and political dispute over inspection authority, fisheries rules, and the handling of a vessel under northern conditions. The immediate facts of that case belong to a specific jurisdictional argument, but its wider importance lies elsewhere. Fisheries enforcement in cold waters rarely fails because nobody knows the rules. It fails, or strains, because the decisive moment arrives fast. A vessel appears during a narrow season, on a productive ground, in conditions where delay changes everything. The window matters more than the statute.

That is the governing difficulty of pulse fisheries. Fish do not always distribute effort evenly across the calendar. Seasonal movements, spawning runs, feeding concentrations, ice retreat, and shifting prey fields can compress economic opportunity into short, intense bursts. For fleets, those bursts reward speed. Regulators punish routine. A licensing system, inspection plan, or observer schedule that looks perfectly adequate over the year may prove threadbare during the two or three weeks when catchability spikes and vessels converge. Time becomes a force in its own right. It orders risk before anyone boards a boat.

This compression changes behaviour at every level. A skipper who might fish cautiously over a long season has an incentive to move harder when the stock is thick, the weather is briefly favourable, and the market or quota position is still open. A processor expects sudden surges in landings and presses for throughput. A port inspector faces a cluster of arrivals that would have been manageable if spread across months. A coast guard or fisheries agency must choose where to be visible and where to accept uncertainty. The same water can therefore generate two calendars: the biological one, which concentrates fish, and the administrative one, which prefers stable workloads. Pulse fisheries are what happens when the first calendar defeats the second.

The concept sounds specialised. It is not. A pulse fishery is simply a fishery in which value is concentrated sharply enough in time that delay carries disproportionate cost. That concentration may arise from spawning aggregation, short feeding windows, ice dynamics, migratory bottlenecks, or regulatory season design. Once it does, the normal logic of compliance changes. Rules still exist, but the incentive to test them increases because each hour matters more. A vessel deciding whether to shift grounds after a bycatch event, whether to wait for clarification at a boundary, or whether to steam for inspection before landing is not merely weighing legality in the abstract. It is weighing the loss of a fleeting opportunity. The clock is on deck.

The northeast Atlantic mackerel disputes after 2010 illustrate one part of this temporal politics, even though they are usually discussed through allocation rather than timing. As mackerel became more seasonally available in Icelandic and Faroese waters, the value of catching during those feeding periods sharpened the dispute over shares. This was not only because more fish were present. It was because they were present when commercial capture was especially meaningful to the fleets involved. A stock that enters a zone for a short, profitable interval creates more pressure than one that remains diffusely available. Seasonal presence can therefore matter politically far beyond its duration. A few decisive weeks may reorder bargaining posture, because those weeks determine who lands first, who fills faster, and who can point to actual utilisation rather than theoretical access.

The same logic appears in demersal fisheries of the north, though it often receives less rhetorical attention. Cod, haddock, capelin, and associated stocks in northern waters are not simply "there" or "not there". They become more or less available in patterns linked to temperature, ice, prey, and spawning behaviour. Fleets organise around those moments. Regulators try to do the same, but institutions are built more slowly than fish move. Patrol assets are finite. Boarding conditions depend on sea state and light. Data from vessel monitoring systems may show concentration, yet translating a cluster on a screen into verified inspection takes time that pulse conditions do not always permit. The result is predictable: during peak pressure, authorities know more than they can act upon.

That gap between awareness and action is the essence of enforcement strain. It is not pure incapacity, still less indifference. It is a mismatch between the tempo of fishing and the tempo of control. Authorities may detect a suspicious pattern, but by the time a patrol vessel reaches the area, the fleet has dispersed or crossed a line. Inspectors may arrive in port just as multiple landings pile up within hours of each other. Electronic reporting systems may flag anomalies faster than analysts can compare, verify, and refer them. Pulse conditions expose every bottleneck in the chain. They show whether a regime has been designed for average workload or for the brief period when ordinary assumptions fail.

A useful example sits in the Bering Sea pollock fishery, especially where bycatch risk from Chinook salmon has turned the season into a more tightly managed contest over timing, avoidance, and continued access. Once bycatch caps and hotspot concerns become operational, vessels no longer respond only to target abundance. They respond to the timing of overlap, the likelihood of moving closures, and the speed with which a haul can change the compliance position of an entire segment of the fishery. Management becomes more temporal than annual statistics suggest. A rule that appears stable on paper may in practice hinge on hours: who enters an area first, who receives avoidance information quickly enough, who continues fishing after the warning signs become visible. Temporal density makes enforcement more forensic and, therefore, more fragile.

This is why pulse fisheries frequently produce compliance gaps that everyone can predict and nobody can fully eliminate. During a compressed season, more vessels fish at once, more catch is landed at once, more transhipment or routing decisions are made at once, and more borderline behaviour becomes tempting precisely because oversight is diluted by volume. None of this requires criminal intent. It is enough that the opportunity cost of delay rises. A captain may remain technically within the rules while pushing them to the limit in space and time. An operator may delay a response, stretch an interpretation, or count on the low probability of immediate checking. Under pulse conditions, even honest systems tend to produce selective enforcement, because not everything can be inspected when everything happens together.

The temptation, then, is to imagine that the answer lies in more surveillance. More satellite coverage. More patrols. More observers. More cameras. These measures can help, sometimes decisively. Yet pulse fisheries expose the limits of surveillance as much as the need for it. Detection does not equal intervention. Data do not inspect holds, sort catches, or calm a jurisdictional dispute on the spot. A satellite track may reveal where vessels crowded during a short aggregation, but it cannot alone establish what was taken, what was discarded, or whether a gear interaction occurred before or after a closure threshold became known. The pressure point is not only visibility. It is response under compression.

A further complication follows from the way rules are often drafted. Many management systems are written against average expected conditions: average seasonal spread, average inspection demand, average landing rhythm, average bycatch probability. Fisheries do not live at the average. They live in peaks. A closure triggered after cumulative evidence may arrive too late if the decisive harm occurs in the first intense days. An observer scheme calibrated for routine coverage may become politically useless when the most disputed behaviour happens during a short window with incomplete placement. A port regime designed for a steady trickle of landings may fail when several vessels arrive almost simultaneously with catches requiring careful verification. Rules that appear sensible in annual review may therefore create a false calm. They govern the middle, while strain accumulates at the edge.

The question is not merely technical. It is constitutional for the fishery. A state that designs around average conditions is, in effect, choosing where uncertainty will be tolerated. It may decide that some peak-period ambiguity is an acceptable price for keeping fleets operating efficiently. Or it may choose stricter peak controls, knowing these will raise costs, generate industry resentment, and perhaps redirect effort into other weakly monitored periods or places. Either way, time is being distributed politically. Governance under pulse conditions is not only about catching up with the fish. It is about deciding whose inconvenience counts.

Still, a caution is needed. Not every short season becomes a governance failure, and not every pulse fishery invites major non-compliance. Some fisheries are highly seasonal yet remain relatively orderly because the rules are simple, participants are limited, and the enforcement chain is trusted enough that a few visible interventions carry weight. Institutional memory matters here. So does fleet structure. A compact fishery with regular participants and predictable landing points is easier to govern under time pressure than a dispersed fishery involving multiple jurisdictions, mixed catches, or intense distributional conflict. Pulse alone does not cause crisis. It amplifies the consequences of whatever weaknesses were already present.

That amplification explains why incident risk rises so quickly when timing, ecology, and politics begin to overlap. A disputed stock arrives early. A bycatch-sensitive species overlaps unexpectedly. A vessel hugs a boundary because the concentration is strongest there. Inspectors are stretched.

Communication with another authority is slow. A routine challenge at sea then acquires a sharper edge because the opportunity at stake is fleeting and the decision cannot easily be revisited next week. Temporal compression turns small frictions into larger signals. A boarding delayed by weather looks, from one angle, prudent. From another, it looks selective. A vessel told to divert for inspection hears not just administration but lost catch. This is how pulse conditions turn ordinary enforcement into a stage for argument.

The quiet motif throughout is the clock. It is visible everywhere once one starts to look. In the sprint to a seasonal aggregation. In the countdown to a quota closure. In the hours between a suspicious track and a possible interception. In the queue of landings waiting for inspection. In the brief interval during which a scientific warning, a radio message, or a management order might still alter conduct before the haul is made. Fisheries law may speak in dates and obligations, but pulse fisheries are governed in minutes and days. The fishery that seems manageable in annual totals may be effectively decided in a weekend.

Designing rules for peak pressure rather than for average conditions therefore sounds sensible, but it carries its own political costs. Peak-oriented regulation is often blunt. It may require pre-emptive closures, denser monitoring, tighter reporting deadlines, stricter move-on rules, or broader discretionary powers for authorities acting under uncertainty. Such tools can reduce harm and strengthen credibility. They can also burden lawful operators, intensify conflict over evidence, and encourage strategic claims that regulators are governing exceptional moments as though they were the whole fishery. The more a regime centres its most restrictive assumptions, the more it risks losing legitimacy among those who experience compliance not as fairness but as permanent suspicion.

This leaves a harder test than it first appears. If fish, vessels, and value now converge in short bursts that overwhelm routine oversight, should governance concentrate power and scrutiny around the peak, accepting higher cost and sharper conflict in order to narrow the gap when it matters most? Or should it preserve broader flexibility and live with the fact that the decisive days of the season are precisely when the rules are least likely to be fully seen, fully checked, or fully believed?

Chapter 3

Uncertainty as Political Currency

How can one survey produce two incompatible demands for caution? That is the quiet paradox at the centre of fisheries politics in remote seas. A sparse dataset, a disputed model choice, a patchy record from observers, and suddenly, restraint is no longer a shared scientific instinct but a weapon used in opposite directions. One actor invokes precaution to freeze access. Another invokes flexibility to avoid surrendering it. The same uncertainty does double duty. It warns, and it bargains.

That is why uncertainty in fisheries governance should not be mistaken for a temporary absence of clarity that better evidence will automatically cure. Sometimes more evidence does narrow the range of honest disagreement. Often it does something rougher. It raises the stakes of interpretation. Once fish move, once survey coverage becomes uneven, once allocation formulas begin to wobble, uncertainty stops behaving like a technical inconvenience. It becomes a medium through which power is exercised. The clipboard starts to matter as much as the patrol vessel.

The central tension lies here: governance needs doubt to remain disciplined, yet politics rewards the selective use of doubt. A responsible regulator cannot pretend to certainty where none exists. A responsible coastal state cannot lightly ignore signs of stock stress or survey weakness. But neither acts in an empty room. Fleets, ministries, processors, and diplomatic counterparts all read uncertainty through consequence. If caution limits a rival's catch, caution can look principled and convenient at once. If the same caution threatens domestic landings or weakens a claim to a larger share, suddenly the language shifts towards adaptive management, incomplete evidence, and the danger of overreaction. The science has not changed. Its political uses have.

This matters because fisheries disputes rarely hinge on biological facts alone. They turn on who gets to treat their own interpretation of incomplete evidence as the reasonable one. A model assumption about productivity is not only a scientific choice when it influences quota size and allocation at the same time. Observer reliability is not merely a technical matter when doubts about it can discredit an entire chain of compliance claims. Survey design is not neutral when a stock's shifting range means some waters are measured more closely than others. Under these conditions, arguments about evidence become arguments about entitlement, and arguments about entitlement quickly harden into judgments about legitimacy.

The stakes are therefore higher than the familiar phrase "data gaps" suggests. Gaps do not simply leave empty spaces. They invite stories. One party says the stock is being undercounted because the surveys no longer reach the right grounds at the right time. Another says the apparent shift reflects temporary conditions and should not unsettle established arrangements. A third insists that uncertainty itself requires tighter limits because the cost of being wrong would be too high. None of these positions is irrational. That is precisely the problem. When several plausible readings exist, institutions must decide whose numbers travel, whose caution counts as prudence rather than opportunism, and whose requests for transparency look like co-operation rather than tactical delay.

Trust enters here in a plain, unsentimental form. It is not goodwill floating above conflict. It is the practical willingness to accept inconvenient information without immediately treating it as manipulation. Shared databases, joint working groups, audit trails, common reporting formats, and agreed review procedures can lower the temperature because they reduce the room for surprise. They also create new pressure points. A party can slow disclosure, contest method, bury disagreement in procedural detail, or release selective fragments that flatter its own case. Transparency helps, but it does not govern itself. The ledger can calm a dispute or deepen it, depending on who controls the pace and framing of what becomes visible.

And yet uncertainty should not be romanticised as permanent intrigue. Many fisheries institutions do manage contested evidence with more steadiness than public debate admits. Scientists revise methods. Agencies compare records. Parties sometimes accept uncomfortable findings because the alternative would cost more in the long run. A bad season of data does not automatically poison a whole regime. This limitation matters. It prevents a lazy cynicism in which every precautionary claim is dismissed as disguised self-interest and every transparency measure as theatre. The harder truth is less dramatic: institutions can absorb uncertainty, but only up to the point where the political rewards of strategic interpretation begin to exceed the reputational cost of being seen to exploit it.

The path ahead follows that strain as it accumulates. It begins with precaution, not as a pure conservation ethic but as a bargaining language that can defend stocks and market position at once. It then moves into the contested ground of missing or uneven evidence, where surveys, models, and observer records stop looking merely technical and start deciding who can make a credible claim. From there, the argument passes into trust, transparency, and reputational pressure, where legitimacy becomes measurable not in sentiment but in whether actors share, verify, and accept information when it hurts.

The decisive question is uncomfortable and unavoidable. When evidence is incomplete but the consequences of acting - or waiting - fall unevenly across fleets and states, what deserves more authority: the cautious interpretation that may protect a stock while preserving an old balance, the adaptive interpretation that may reflect changed reality while inviting opportunism, or the institutional process that asks everyone to live for a while with uncertainty none of them truly trusts?

The politics of precaution

In August 2013, the European Union imposed trade measures against the Faroe Islands during the dispute over Atlanto-Scandian herring and north-east Atlantic mackerel. The public justification turned on orderly management and the need to prevent overfishing. The political effect was sharper. Precaution had become an instrument in a struggle over who would bear the costs of ecological change and who would be allowed to convert new access into durable advantage. A cautious rule can conserve a stock. It can also freeze a market position. The same language does both.

That dual function is the point from which any serious discussion of precaution in fisheries must begin. Precaution is often presented as a moral and technical response to uncertainty: if the stock may be at risk, managers should restrain catches rather than gamble on optimistic assumptions. That principle remains indispensable. It has saved fisheries from the fantasy that absence of proof is proof of abundance. Yet in contested waters, and especially where stocks move across jurisdictions, precaution is never merely a conservation ethic. It is also a bargaining language. Actors invoke caution when caution protects their interests, and invoke flexibility when flexibility protects them instead. The word does not lie. It simply carries more politics than it first admits.

This becomes clearest when a stock's distribution changes faster than the governing bargain. A state that has long benefited from an established allocation may suddenly discover a passionate attachment to conservative interpretation: the survey record is incomplete, the apparent redistribution may be temporary, the prudent course is to avoid destabilising the system. A gaining state, by contrast, may insist that rigid caution now serves only to preserve an outdated division of benefits. The same parties might reverse those positions in another fishery. That is why precaution must be read not only as a principle but as a situational claim. It tells us something about the stock. It also tells us who fears adjustment.

The structure of the argument is usually familiar. Show uncertainty, name risk, explain the need for restraint, and then smuggle in a distributive consequence. A government points to incomplete survey coverage or shifting productivity and argues that catch expansion would be irresponsible. The scientific premise may be sound. The political implication is rarely neutral. If the resulting restraint falls hardest on newer entrants or on states whose present access has improved, precaution functions as a gatekeeping device. It preserves the value of older arrangements by translating ecological doubt into institutional delay. This does not make the claim insincere. It means sincerity and advantage can coincide.

The mackerel disputes of the late 2000s and early 2010s make this pattern unusually legible. As the stock became more available in Icelandic and Faroese waters, those states argued that the existing allocation no longer reflected ecological reality. Their counterparts did not simply deny the movement. They framed the dispute through responsible management, warning against unilateral catches and stressing the dangers of exceeding scientifically advised totals. That language carried real force. Fisheries can be damaged by competitive escalation masked as belated fairness. But the same language also defended a structure from which established participants had long benefited. Precaution, in this setting, was not a mask for politics. It *was* politics in a respectable form.

This is why fisheries negotiations often feel morally crowded. Few actors arrive saying openly that they wish to preserve market share or quota advantage under the banner of self-interest. They arrive speaking of stewardship, sustainability, scientific prudence, and institutional discipline. Much of that language is justified. The trouble is that identical words can serve opposite strategies. One side says caution because the stock is uncertain; another hears caution because the share is threatened. One side says flexibility because the fish have moved; another hears flexibility because someone wants to convert a temporary gain into permanent entitlement. A closed door matters here. Much of the real work of precaution occurs not in public declarations, but in the bargaining room where actors decide how much uncertainty must count before the rules move.

The politics sharpen further because precaution rarely distributes pain evenly. A lower catch recommendation may hurt all parties in theory, but not equally in practice. A state with diversified fisheries, strong processors, and stable market access can absorb restraint more comfortably than a state or region whose recent investments depend on a newly available stock. A government facing coastal unemployment will hear "caution" differently from one defending a mature export position. The abstraction of risk becomes concrete in a decision: whether to keep a plant open on a reduced season, whether to support a unilateral quota, whether to threaten trade measures, whether to accept an interim arrangement that conserves the stock but leaves altered access largely unrecognised. Precaution lives inside those choices. It is never only a sentence in an advisory document.

The Faroe Islands dispute in 2013 showed how quickly precautionary language can travel from fisheries management into market pressure. The European Union did not need to turn the argument into a naval drama. It could instead use trade restrictions to present its stance as a defence of orderly, sustainable management against excessive catching. That move signalled two things at once. It signalled that scientific and regulatory discipline mattered. It also signalled that market access could be used to enforce a preferred interpretation of prudence. Once precaution acquires downstream instruments such as import restrictions, certification pressure, or port access constraints, it becomes more than a technical threshold. It becomes a means of disciplining rival claims to the resource.

The concept, therefore, cannot be understood adequately if it is confined to stock biology alone. Precaution functions institutionally through rules about total allowable catch, timing, bycatch limits, observer coverage, and evidentiary standards. It functions economically through those who can land, sell, insure, and process the catch under conditions of uncertainty. It functions diplomatically through whose interpretation of risk becomes the default position in negotiation. A state that succeeds in making its caution appear normal and responsible often gains the upper hand before any quota numbers are finalised. The battle is partly semantic. Whoever defines prudence defines the burden of proof.

That burden of proof is never fixed. In some disputes, the side urging expansion or revision must prove beyond reasonable doubt that the ecological shift is durable and significant. In others, the side defending the old bargain must prove that delay will not amount to a political confiscation of new access. These are different procedural worlds. The first favours incumbents. The second favours claimants. Precaution sits between them, often without being named as such. It answers a prior question: who should bear the risk of acting too soon, and who should bear the risk of waiting too long? Once phrased that way, the politics becomes unavoidable.

A useful contrast appears in the long management of shared Barents Sea stocks through the Joint Norwegian-Russian Fisheries Commission. The arrangement is not free of conflict, nor should it be romanticised. What it demonstrates, however, is that precaution need not always operate as a blunt distributive weapon if the parties accept ongoing procedures for scientific review, quota negotiation, and mutual adjustment. In such settings, caution can serve more as a shared discipline than as a tactical veto. That distinction matters. It shows that the political use of precaution is not inevitable in the same way everywhere. Institutions can soften it. They cannot abolish it.

Still, even comparatively durable institutions do not eliminate asymmetry. A party that expects future access gains may favour provisional flexibility now, while a party more exposed to downside ecological risk may insist on tighter restraint. Both positions can be defended honestly. The line between principled caution and strategic caution is rarely bright. That is precisely why reputational politics matters so much. A government seen to weaponise precaution too transparently may pay a diplomatic cost later when it seeks deference in another dispute. A government seen to dismiss caution whenever its own fleets stand to gain may damage its credibility with scientists, partners, and markets. Reputation becomes a form of ballast. It cannot stop self-interest, but it can slow the most cynical uses of scientific uncertainty.

This is also why appeals to "follow the science" often disappoint in fisheries politics. Science can tell managers a good deal about stock condition, distribution, and the consequences of different catch levels. It cannot decide how much asymmetrical risk different parties should absorb while uncertainty persists. Nor can it decide whether institutional stability deserves temporary priority over distributive revision, or vice versa. Those are political judgments. Precaution enters as the grammar through which they are argued. To say "we must be cautious" is often to say something more specific and more interested: we should keep the current arrangement until evidence compels change, or we should loosen it because evidence already compels change. The science informs the sentence. It does not finish it.

The politics of precaution becomes especially visible where competitors alternate between caution and flexibility across issues without embarrassment. A government may demand strict interpretation for total catches while urging relaxed treatment of zonal attachment. An industry group may call for conservative assumptions about a rival fleet's claims while embracing adaptive reasoning for its own access. These reversals are not necessarily hypocritical in a crude sense. Fisheries are genuinely multidimensional, and different forms of uncertainty justify different responses. But the pattern should still make us wary of treating precaution as an innocent fixed principle. In practice it behaves more like a repertoire. Actors choose from it.

A credible limitation must be admitted. Not every appeal to precaution hides a strategic motive, and not every call for flexibility signals opportunism. Stocks can collapse under rosy assumptions. Scientific uncertainty can be grave enough that restraint is plainly warranted, whatever the distributional consequences. Likewise, rigid caution can sometimes ossify a management system so badly that it ceases to reflect the fishery it governs. The presence of politics does not cancel the reality of ecological danger. It means only that ecological danger is interpreted through institutions whose participants must answer to fleets, ports, processors, markets, and voters as well as to scientific advice.

That limitation sharpens rather than weakens the argument. Because precaution can be both sincere and strategic, the key analytical task is not to unmask every cautious position as self-serving. It is to ask what the cautious position does. Does it reduce a genuine conservation risk while preserving room for later adjustment? Does it lock in an incumbent advantage by converting uncertainty into endless postponement? Does flexibility accommodate a real shift in stock distribution, or does it invite a competitive rush that makes future co-operation harder? The measure lies in behaviour and consequence, not in declared virtue. Watch what the rule protects. Watch who must wait.

The quieter truth is that fisheries governance depends on precaution precisely because uncertainty cannot be eliminated. Stocks move, models strain, survey coverage remains incomplete, and institutional trust is always partial. The question is never whether to act under uncertainty. It is whose uncertainty counts, whose losses are treated as intolerable, and whose claims must clear the highest evidentiary hurdle before policy will move. Precaution is not outside that struggle. It is one of its most polished forms.

The next time a state, ministry, or regional body calls for caution, the serious question will not be whether caution sounds wise. It usually does. The harder question is what kind of future that caution is purchasing: a genuine buffer against ecological error, or a temporary conservation argument that also happens to keep a rival fleet out of the water just long enough for an old advantage to harden again.

Data gaps and contestable evidence

In 2010, Iceland set a unilateral mackerel quota after the stock had become far more available in Icelandic waters during the summer feeding season. The dispute that followed across the north-east Atlantic did not rest on politics alone. It rested on a harder and more durable problem: the fish were visibly present, but the evidentiary system used to describe the stock had been built around an older geography. Iceland could point to catches and local availability. Others could point to historical participation, established survey routines, and the dangers of turning a recent shift into a permanent entitlement. The same fish supported different stories. The chart was not enough.

That is how data gaps work in remote seas. They rarely appear as blank spaces waiting patiently to be filled. They appear as disagreement over what kind of evidence counts, which absences matter, and how much uncertainty a management system can bear before it begins to misdescribe the fishery it claims to govern. In a stable setting, missing information may remain a technical inconvenience. In a moving one, it becomes political currency. A stock enters waters not well covered by earlier surveys. A fleet reports new concentrations. A scientific body adjusts a model cautiously. A coastal state argues that caution is now biased by another name. The ruler wobbles.

The important point is not that fisheries science becomes unreliable when conditions change. It is that evidence becomes easier to contest when survey design, sampling windows, and model assumptions lag behind ecological reality. A research cruise follows the route that has long yielded a useful picture of the stock, but the stock has stretched beyond that route, or shifted deeper, or changed its seasonal residence in ways the old design captures imperfectly. A country with newly improved access sees undercounting.

A country defending an older share sees temporary fluctuation being overread. Both positions can draw strength from the same underlying fact: the observational system no longer fits the sea as neatly as it once did. That is not only a scientific strain. It is a distributive opening.

Survey design often looks neutral from the outside because it arrives dressed as method. It involves transects, timing, vessel availability, acoustic assumptions, trawl verification, and historical comparability. But design choices have consequences far beyond the laboratory. When a survey happens in one season rather than another, it may privilege one stage of the stock's annual movement over another. When it excludes newly important waters because access is difficult, expensive, or politically constrained, it may preserve the authority of a dataset while weakening its descriptive honesty. When it extends to new grounds, it may produce a different picture that unsettles existing rights. Methods matter politically because they determine what can be seen, and what can be seen enters directly into the arguments over total allowable catch and quota share.

The mackerel conflicts involving Iceland and the Faroe Islands made this painfully clear. Coastal states could acknowledge the stock's wider distribution without agreeing on the significance of that shift. Was summer presence in new waters enough to justify major reallocation? Did feeding distribution deserve the same weight as spawning geography? How much of a recent pattern had to be observed before it counted as a new baseline rather than a temporary excursion? These were not questions science could avoid, yet neither were they questions science could answer by measurement alone. Evidence could describe where the fish were encountered and in what quantities. It could not decide which part of that annual behaviour should carry the heaviest political weight. Once that gap opens, contestable evidence becomes unavoidable.

A second example sits in the Central Arctic Ocean, where the Agreement to Prevent Unregulated High Seas Fisheries was signed in Ilulissat in October 2018. The agreement is often praised as a rare act of foresight, and rightly so. But its logic also illuminates the problem of missing evidence. The central Arctic high seas were not then the site of an active commercial fishery. Scientific understanding of potential future stocks, productivity, and distribution remained limited, and the physical environment made comprehensive observation difficult. Rather than pretend those limits did not matter, the parties treated the absence of settled evidence as grounds for restraint and joint research. That response was prudent. It also reveals a broader truth. Where no party can convincingly claim a complete picture, the politics of evidence shifts from arguing over whose dataset is best to arguing over who should be allowed to act before the picture improves. Even ignorance has distributional consequences.

Remote seas generate these disputes because data do not arrive evenly. Survey vessels follow national budgets, ice conditions, diplomatic permissions, and practical priorities. Observer coverage depends on fleet structure, regulatory design, cost, safety, and political acceptance. Logbook records vary in quality and credibility. Electronic systems generate huge streams of information, but much of it concerns movement rather than catch composition, stock identity, or the biological meaning of presence. The result is a layered evidentiary field in which different actors know different things with different levels of confidence. Scientists may know the assessment model is being pushed beyond some of its older assumptions. Fishers may know the stock is turning up where the model barely looks. Ministries may know that acknowledging either point too fully could damage their bargaining position. The data gap is not just in the sea. It is also between institutions.

Observer reliability becomes especially fraught under such conditions. In theory, observers convert uncertain deck activity into credible records. In practice, their authority depends on who appoints them, where they can be placed, what they can safely see, how their observations are recorded, and whether anyone downstream trusts the chain of custody. In remote fisheries, observer placement may be incomplete because seasons are compressed, vessels are numerous, or conditions are hazardous. One party then treats observer data as indispensable evidence of bycatch, effort, or catch composition. Another points to patchy deployment, limited sample size, or differences in training and says the record cannot bear the weight being placed on it. Neither side is necessarily inventing the problem. The same observational weakness that invites prudence also invites narrative warfare.

Model choice compounds the difficulty because models do not simply process data; they arrange the relationship between data and judgement. One model may place greater weight on long time series and therefore smooth recent anomalies. Another may respond more readily to apparent environmental change but at the cost of greater sensitivity to short-run noise. One may embed assumptions about catchability or recruitment that fit older conditions better than present ones. Another may incorporate broader spatial signals while weakening comparability with past advice. When parties contest model choice, they are contesting more than technique. They are contesting which uncertainty deserves institutional authority. A model that looks conservative to one side may look stubbornly archaic to another. A model that looks adaptive may look opportunistically loose.

This is why "follow the science" often fails as a political instruction in fisheries disputes. The phrase implies a singular path where several scientifically defensible paths may coexist, especially in transition periods. Scientific bodies can narrow the field, exclude obviously weak claims, and clarify the consequences of different assumptions. They can do a great deal. What they cannot do is erase the normative choice hidden inside evidentiary preference. Should managers favour continuity in the face of noisy signals, knowing that this may entrench older shares? Or should they give greater weight to recent distributional evidence, knowing that this may harden temporary gains into new precedent? The answer cannot be read directly from the spreadsheet. Someone must decide how much instability the system is willing to tolerate.

The ruler wobbles again here, and it wobbles most when total allowable catch and allocation are contested at the same time. If uncertainty concerned only stock size, the problem would already be serious. But when the same uncertainty also influences who is entitled to benefit, every methodological dispute acquires a double edge. A lower catch recommendation can intensify competition over shares. A revised estimate of spatial distribution can strengthen claims for reallocation. A disagreement over survey coverage can therefore affect both conservation and distribution at once. That is why debates that sound technical so often become emotionally charged. The numbers do not only govern fish. They govern winners.

A further difficulty arises from the difference between abundance and accessibility. A survey may indicate that a stock remains broadly healthy, while a fleet experiences sharp local changes in catchability because fish are arriving earlier, moving differently, or concentrating in newly relevant grounds. A coastal state can then argue, with some force, that the scientific advice is not wrong but incomplete for the purpose that now matters most: allocation under changed access. The other side replies that access should not be mistaken for stock-wide entitlement. Both are making an evidentiary claim. One privileges operational reality. The other privileges systemic caution. Here again, whose numbers travel depends not only on scientific merit but on institutional rules about what kind of evidence is allowed to matter.

It is tempting to think that more transparency would solve much of this. Shared databases, open model documentation, joint working groups, and common audit procedures can indeed reduce suspicion. When parties can inspect survey methods, compare raw data, and follow model revisions openly, room for outright fantasy narrows. But transparency is not a neutral solvent. It creates new grounds for obstruction and selective emphasis. An actor may release data slowly, overwhelm a counterpart with technical detail, publicise only the fragment most helpful to its own case, or use procedural demands for openness as a way to delay unwanted decisions. Transparency improves the quality of contestation more often than it eliminates contestation. That is still an achievement, but it is a narrower one than reformers sometimes imagine.

A credible limitation should be acknowledged. Not every evidentiary dispute reflects strategic behaviour, and not every data gap becomes politically explosive. Some remote fisheries continue to function because institutions have built habits of joint interpretation strong enough to hold disagreement within procedural bounds. The Joint Norwegian-Russian Fisheries Commission, for all its tensions and historical baggage, demonstrates that shared scientific work can create a stabilising structure in which disputed evidence does not automatically trigger public rupture. Yet even there, co-operation works because parties accept not only the value of better information but the authority of a process that allows no single dataset to become sovereign by national decree. The lesson is reassuring and sobering at once. Good institutions can make evidence travel more calmly. They cannot make uncertainty disappear.

The real issue, then, is institutional carriage. Whose numbers travel? Not whose numbers are numerically larger, but whose numbers survive the journey from survey vessel to scientific body, from scientific body to negotiating table, from negotiating table to quota decision, and from quota decision to accepted fact on the water. Data do not travel by themselves. They move through committees, models, national agencies, diplomatic routines, and reputational filters. A dataset gathered by one state may be treated with caution if that state is seen as too exposed to the distributional consequences. An observer report may be doubted because earlier reports were contested. A revised model may be correct and still fail politically because the timing of its correction is too convenient for some actors to trust. Legitimacy is the vehicle.

That is why contested evidence matters so much in Arctic and sub-Arctic fisheries. It is not merely that knowledge is hard to gather. It is that incomplete knowledge enters a system in which fisheries are valuable, mobile, and deeply politicised. Under those conditions, every evidentiary weakness can be used to defend continuity or demand revision. Every data gap is both a scientific problem and an invitation to narrative. Some actors will speak in the language of prudence, others in the language of reality, others in the language of fairness. All three may rest on genuine observations. The difficulty lies in deciding which observations acquire governing force.

The practical question that follows is sharper than it first appears. When remote seas yield uneven surveys, disputed observer records, and model choices that rearrange both conservation advice and quota claims, what should count as sufficient evidence for action: the dataset most comparable to the past, the one most responsive to present change, or the one carried by institutions credible enough that even disappointed parties can still live with its conclusions?

Trust, transparency, and reputational leverage

In October 2018, in Ilulissat, Greenland, ten parties signed the Agreement to Prevent Unregulated High Seas Fisheries in the Central Arctic Ocean. The immediate result was not a dramatic act of interdiction at sea, but a quieter achievement: parties that could not yet police every future vessel movement across those waters agreed to make restraint, data-sharing, and joint scientific work part of the governing architecture before a commercial fishery took hold. That choice revealed something easy to overlook in the language of enforcement. Order does not depend only on ships, aircraft, and legal authority. Sometimes it depends on whether states believe one another's records enough to postpone a race. The ledger can hold a line.

This is the practical importance of trust in fisheries governance. Not trust as sentiment, still less as diplomatic warmth, but trust as the working assumption that information supplied by another party is sufficiently reliable to justify restraint in one's own conduct. A patrol vessel can compel compliance in a narrow sense, for a moment, against a particular target. Trust works differently. It widens the space in which actors accept verification without immediate confrontation. It allows an observer report, a shared stock assessment, a landing declaration, or a vessel track to do political work that force alone cannot do cheaply or continuously. In distant seas, where presence is always partial, that kind of credibility becomes an operational asset.

The point becomes clearer once one admits the obvious: most fisheries are governed with far less direct inspection than public rhetoric implies. Authorities do not watch every haul, verify every species mix in real time, or board every vessel whose behaviour raises questions. They rely instead on chains of information - logbooks, monitoring systems, port records, observer notes, scientific surveys, licence data, bilateral exchanges, regional working groups. These systems function only if participants treat some information as presumptively usable. Once that presumption breaks down, enforcement becomes both more expensive and more political. Every discrepancy looks intentional. Every omission looks strategic. Every request for more time to validate data sounds like a prelude to evasion.

That is why transparency mechanisms often substitute for weak enforcement, at least for a while. Shared databases, routine scientific exchanges, standardised reporting formats, audit trails, joint commissions, cross-checks between flag states and port states, and agreed procedures for reviewing anomalies all reduce the need for constant coercive presence. They do so not by eliminating conflict, but by changing its terms. A dispute that might otherwise erupt as an accusation can be routed into documentation. A suspicious landing can be compared against prior declarations. A claim about stock distribution can be tested against shared survey work rather than asserted as national truth. Transparency, in this sense, is not a moral flourish. It is an administrative way of turning uncertainty into something less combustible.

The long record of the Joint Norwegian-Russian Fisheries Commission illustrates this better than any slogan could. Established in 1976, the commission has endured periods of broader geopolitical tension precisely because it institutionalised extensive routine scientific and managerial exchange around shared northern stocks. That history should not be romanticised. Disagreement has never vanished, and broader political relations have always mattered. But the arrangement demonstrates a crucial fact: when parties repeatedly compare assessments, negotiate catches, and maintain working-level contact, they create a stock of institutional credibility that can absorb shocks. A shared survey or agreed stock estimate does not become unchallengeable. It becomes harder to dismiss casually, because doing so imposes a reputational cost on the dismissing party.

Reputational leverage is often misunderstood because it sounds soft. In practice, it has visible effects. A state seen as manipulating data, ignoring agreed procedures, or disclosing information selectively may find it harder to persuade others in later negotiations, harder to defend its fleets in a contested incident, harder to secure market confidence, and harder to frame its own caution as principled rather than tactical. Reputations travel through ministries, regional bodies, scientific networks, buyers, and insurers. They affect whether a request for flexibility is heard as a reasonable adaptation or as opportunism dressed in technical language. In this domain, legitimacy is not decorative. It changes the burden of proof.

The northeast Atlantic mackerel disputes after 2010 provide a sharp counterpoint. As the stock's summer distribution shifted more strongly into Icelandic and Faroese waters, the dispute over quota shares quickly became a dispute over whose evidence and whose procedures deserved authority. Established participants stressed scientific discipline, orderly management, and the danger of unilateralism. Iceland and the Faroe Islands stressed the changed ecological reality and the unfairness of a system that lagged behind it. Publicly, both sides invoked responsibility. Beneath that language sat a harder contest over credibility. Could newly assertive participants persuade others that their claims reflected durable stock behaviour rather than opportunistic interpretation? Could established parties defend the existing order without looking as though they were using procedure to freeze advantage? That conflict was never only about fish. It was also about which actors could make their account of the fishery travel.

This is where transparency helps, but only conditionally. Open methods, shared records, and common reporting standards can narrow the space for fantasy. They can expose weak claims earlier and reduce the temptation to dramatise every disagreement. Yet transparency does not float above politics. It creates its own arenas of manoeuvre. A party can disclose data slowly. It can flood a counterpart with technical detail while withholding interpretive context. It can insist on procedural purity when delay serves its interests, then demand pragmatic flexibility when time begins to hurt. A shared database can therefore reduce suspicion and generate new forms of obstruction at once. The same ledger that builds trust can become a battlefield over timing, completeness, and framing.

Timing matters more than reformers sometimes admit. A dataset released late in a negotiation season may be formally transparent yet politically manipulative. An audit trail may exist but remain so fragmented across agencies that it cannot settle a live dispute before the fishing window closes. A scientific working group may function well in general while becoming paralysed precisely when a revised assessment would alter quota claims or invite domestic backlash. Transparency, then, is not only about what is shared. It is about when it is shared, in what form, and with what expectation of use. Information that arrives too late to discipline behaviour often serves rhetoric better than governance.

Still, one should resist easy cynicism. Transparency mechanisms often do work. Shared scientific assessments can prevent disputes from hardening prematurely. Electronic traceability can deter obvious laundering. Common vessel records and reciprocal inspection arrangements can give states more confidence in one another's follow-through. Routine publication of measures and outcomes can help smaller actors, including researchers and civil society, identify where official claims do not match practice. These gains are real. They matter especially in remote fisheries where direct enforcement capacity is thin. But they matter because they alter incentives, not because they make interests disappear. Actors remain strategic under transparent conditions. They are simply strategic in narrower corridors.

A useful limitation must be acknowledged. Trust can substitute for weak enforcement only up to a point. It cannot rescue a regime in which incentives to cheat are overwhelming, sanctions are meaningless, or ecological and political change has made the underlying bargain too contested to sustain by record-sharing alone. A fleet that believes detection will not lead to real consequences may treat transparency as a performance rather than a discipline. A state facing acute domestic pressure may abandon cooperative routines despite years of prior credibility. Reputational leverage also works unevenly. It bites hardest where actors care about future access, diplomatic standing, scientific standing, or market recognition. It matters less where alternatives exist and the immediate prize is large enough to justify reputational loss.

This is why legitimacy should be treated as a governance asset but not as magic. It sits alongside patrol hours, port controls, legal authority, and administrative competence. It does not replace them. What it does do is change the cost of using them. A coast guard action backed by trusted documentation looks different from the same action taken in an atmosphere of mutual suspicion. A port denial grounded in a credible audit trail is easier to defend than one resting on opaque allegations. A state that has built a reputation for reliable reporting can often secure patience in moments of uncertainty that a distrusted counterpart will never receive. Legitimacy lowers transaction costs. It expands procedural room. It can delay escalation long enough for slower instruments of governance to work.

The problem, of course, is that legitimacy accumulates slowly and can be spent quickly. Years of competent data-sharing can be damaged by one conspicuous failure to disclose, one politicised assessment, one high-profile enforcement inconsistency, or one perception that technical procedure is being used to disguise distributive self-interest. Fisheries governance suffers especially from this fragility because the same information systems that support co-operation also record its breach. A missing dataset, a late notification, a discrepancy between landing records and declared catch, or a refusal to subject a national estimate to joint review can become symbolic events. The ledger remembers.

That memory gives reputational leverage its force. It also makes recovery difficult. Once a party is widely seen as selective with evidence or procedural in bad faith, even accurate future claims may travel poorly. Counterparts will demand more proof, more auditing, more independent validation, more formalisation of what previously rested on working trust. Enforcement burdens rise as a direct consequence. What could once be managed by exchange and routine cross-check now requires more intrusive verification. In this sense, transparency does not merely supplement enforcement. It can prevent the inflation of enforcement costs by preserving a zone of credible self-reporting and mutual recognition.

The institutional question underneath all of this is simple and severe: whose records can others afford to rely on? Not whose records are flawless - none are - but whose records are good enough, consistent enough, and auditable enough that relying on them does not look reckless. That judgment shapes everything from quota negotiation to incident response. It determines whether an ambiguous vessel track becomes a conversation or a confrontation, whether a revised stock assessment becomes a basis for interim adjustment or a trigger for nationalist rhetoric, whether a compliance dispute stays inside technical channels or spills into public accusation. In remote seas, where perfect oversight is impossible, this question may matter as much as the formal rule itself.

The deeper implication is that transparency is not simply about seeing more. It is about making what is seen governable by others. Data that cannot be trusted, compared, or carried across institutions does little to reduce uncertainty. It may even widen it. The same applies to reputation. A good reputation is not a moral halo. It is the accumulated effect of records, procedures, disclosures, and follow-through that make future claims cheaper for others to believe. In a fisheries frontier defined by distance and moving stocks, that cheapness is precious.

The unresolved test is therefore not whether transparency is good in the abstract. It almost always is. The harder question is when transparency genuinely strengthens shared restraint, and when it merely furnishes new tools for delay, framing, and selective exposure. If legitimacy is as real a governance asset as patrol hours, how should states act when preserving that asset requires disclosure and procedural patience, yet the immediate political reward lies in withholding, dramatizing, or publishing only enough of the ledger to win the season?

Chapter 4

Quota Allocation and Bargaining

Who owns a fish that has moved? The question sounds almost childish until a stock shifts across a boundary, a port loses landings, a fleet claims new access, and a government discovers that biology has reopened what law had tried to settle. Then the childishness vanishes. The issue is no longer how many fish the sea can bear, but which principle will decide who may take them, land them, sell them, and speak for them. The ledger changes hands.

That is why quota allocation sits at the centre of fisheries politics. Conservation can tell us whether a fishery might remain biologically viable under restraint. Allocation asks something harder: how the value of that restraint, or of that opportunity, should be divided among actors with different histories, different geographies, and different levels of power. The total catch may be perfectly defensible in scientific terms and still provoke bitter dispute because sustainability does not settle entitlement. A fishery can be healthy and politically unstable at the same time. The arithmetic of distribution is where most of the real conflict begins.

The tension grows sharper in the Arctic and sub-Arctic because none of the available allocation principles is neutral. Historical catch rewards those who entered early, invested early, and secured recognition under older ecological conditions. Zonal attachment promises a more contemporaneous logic, yet quickly becomes slippery when fish move by season, depth, and life stage rather than by neat annual residence. Needs-based claims carry moral force where food, employment, or peripheral communities depend heavily on the fishery, but those same claims can look suspiciously elastic when deployed in negotiation. Every principle sounds principled when one benefits from it. Every principle reveals its distributive bias when one does not.

This is the deeper reason quota bargaining rarely ends with a tidy formula. Allocation systems do more than divide catch. They shape incentives.

They decide whether fleets race to establish a presence that might later strengthen a claim, whether officials defend older baselines because revision would devalue long-standing investments, whether fishers discard or underreport when the species mix undermines their quota position, whether a coastal community feels represented by the system that speaks in its name. A quota share is never only a number. It is an expectation regarding future access, bargaining power, and political recognition. Once that is understood, it becomes obvious why allocation arguments persist long after the conservation argument appears to have been settled.

The stakes are not confined to the negotiating table. They reach ashore into leasing markets, vessel finance, processing plants, regional labour politics, and the symbolic life of particular species. A quota that becomes transferable may improve economic efficiency while loosening the tie between resource and place. Ownership drifts. Leasing concentrates. Communities that still see the fish as theirs watch rights migrate into distant balance sheets. In another setting, a government may stabilise a bargain through side concessions - access swaps, joint ventures, scientific co-operation - only to discover that such pragmatism looks dishonourable once it passes through domestic debate. Allocation is therefore not simply a technical branch of fisheries management. It is where political economy, legal principle, and public legitimacy collide.

That collision matters because it determines whether the system rewards patience or gamesmanship. When future quota share depends too heavily on recent catch, operators may rush to maximise presence rather than fish cleanly or cautiously. When a stock's location changes and the governing formula lags behind, states may prefer unilateral announcements to prolonged procedural restraint. When quota becomes a tradable asset detached from the harbour, the fishery may look more efficient on paper while becoming less defensible in democratic terms. These are not side effects. They are the political life of allocation. A regime that ignores them can preserve order for a while, but only by storing resentment for later release.

Still, not every bargain over quota is doomed to harden into grievance. Some systems endure because they link principles rather than worship one of them. They combine historical participation with current distribution, economic flexibility with social safeguards, or numerical shares with compensating arrangements that spread adjustment costs more tolerably. That limitation deserves attention. It prevents the familiar slide into fatalism. Bargaining is not doomed merely because interests diverge. It fails when institutions cannot absorb the distributive meaning of ecological change, or when domestic politics turns every compromise into evidence of betrayal. The difficulty lies less in the existence of trade-offs than in the refusal to name them honestly.

The path ahead follows those trade-offs where they bite hardest. It begins with the rival logics of allocation itself and the kinds of behaviour each one rewards or punishes. From there the argument moves into the appeal and instability of tying shares to where fish appear in a warming sea, then into the market life of quota once it becomes transferable, leaseable, and prone to concentration. The pressure sharpens through waste, racing, and strategic reporting, then broadens into bargains built from linked concessions rather than clean arithmetic, before meeting the domestic political force of species that come to stand for far more than protein or price.

Everything that follows turns on a stubborn fact. A quota system does not merely distribute fish; it distributes legitimacy. It tells fleets what kind of conduct is worth the risk, tells communities whether their dependence counts, tells governments what sort of compromise they can survive, and tells rivals which past the present is supposed to honour. When the next stock shift and the old formula start to look partial, what should weigh more heavily: the history written into prior catches, the geography written into present distribution, or the political need to keep the bargain believable enough that no one feels compelled to test it at sea before it can be revised on paper?

Allocation principles and their trade-offs

In 2010, Iceland set a unilateral mackerel quota after the stock had become markedly more available in Icelandic waters during the summer feeding season. The immediate dispute that followed with the European Union, Norway, and others was often described as a quarrel over science and co-operation. It was certainly both. But at its core it was a quarrel over principle. Should quota follow historical catch, rewarding those who had built fleets and markets under the old distribution? Or should it follow the fish, rewarding states that could now point to a visible, commercially significant presence in their own waters? The stock moved. The argument stayed.

That episode remains so instructive because it exposed the central fact of fisheries politics without disguise. A sustainable total catch does not tell anyone how the catch should be divided. Biology can discipline the size of the pie. It cannot decide who deserves the larger slice. That second question is where grievance accumulates, because every plausible allocation principle carries a different moral story, a different economic effect, and a different invitation to strategic behaviour. The principle does not merely distribute fish. It distributes patience.

Historical catch is often the oldest and most stubborn of these principles because it rests on a compelling intuition. Those who have long participated in a fishery, invested in vessels and processing, accepted management disciplines, and built communities around the resource should not be displaced lightly by short-term ecological fluctuation or late-arriving claimants. That argument has weight. It protects reliance. It rewards continuity. It stabilises expectations in a sector where capital costs are high and seasons are uncertain. The show is familiar: a fleet with decades of participation, a port organised around predictable landings, a scientific and regulatory system calibrated to a long pattern of use. The name for the governing logic is historical entitlement. The explanation is simple enough. Past access becomes a warrant for future share. The implication is more troubling. History, once converted into quota, can look less like earned continuity than like a perpetual mortgage on the future.

That is why historical catch so often appears fair to incumbents and exclusionary to everyone else. It favours those who were already present when the stock was abundant or politically accessible. It tends to underweight late ecological shifts, especially where newer coastal states can demonstrate current presence but not a long paper trail of landings. It can also reward an older era of management, including periods in which some actors enjoyed access under conditions that would now be politically or ecologically unacceptable. Historical catch therefore stabilises one kind of order while breeding a second kind of resentment. A claimant who sees fish in local waters but receives only a token share does not experience the principle as a neutral continuity. It experiences it as a frozen advantage.

This is where zonal attachment enters with its own apparent clarity. If fish are found in a state's exclusive economic zone, surely that state should receive a corresponding share. The attraction of the principle lies in its modernity. It sounds empirical rather than inherited. It promises to align rights with current ecological reality rather than with older deals. In a warming Arctic and sub-Arctic, where stocks redistribute, that promise becomes politically potent. A coastal state can point to the fish and say, in effect, that the law should stop pretending they are elsewhere. The principle names this claim as spatial fairness. The explanation is that location ought to matter. The implication is less tidy than the slogan suggests. Where, exactly, are the fish - and when?

Zonal attachment quickly becomes contentious because distribution is not static. Stocks move by season, by depth, by life stage, and sometimes by feeding opportunity rather than durable residence. A fish may spend part of the year in one zone, spawn in another, and be most commercially accessible in a third. Measure annual biomass and one result appears. Measure feeding-season concentration and another emerges. Measure spawning grounds, juvenile abundance, or catchability and yet another story takes shape. "Where the fish are" turns out not to be a single fact but a contested measurement choice. That does not make zonal attachment useless. It makes it political. The principle appears objective until parties start deciding which season, which distribution, and which stock component deserves the greatest weight. Then its neutrality vanishes.

The mackerel disputes after 2010 showed exactly this difficulty. Iceland and the Faroe Islands argued that the stock's increased presence in their waters justified much larger shares. Their counterparts acknowledged distributional change to varying degrees but resisted what they saw as an attempt to transform feeding-season availability into a full redistribution of entitlement. Both sides spoke a recognisable language of fairness. Both also chose the evidentiary frame most helpful to their claim. Zonal attachment did not resolve the dispute because the principle itself depended on prior choices about what kind of presence mattered most. The fish in the water were real. So was the interpretive struggle over which fish counted.

Needs-based claims add a different moral vocabulary. These arguments do not ask only who has fished historically or where the stock is found. They ask who depends on the fishery most heavily for food security, employment, or regional stability. A smaller coastal state, a remote community, or a politically exposed region may claim that formal equality is not substantive fairness where livelihoods hinge more directly on the resource. Such claims can be deeply persuasive, especially where alternative employment is scarce and fisheries carry cultural as well as economic meaning. The relevant name here is distributive justice. The explanation is that equal treatment can entrench unequal vulnerability. The implication is complicated. Needs-based reasoning invites empathy and policy flexibility, but it also opens the door to elastic definitions of need and to bargaining positions that are morally resonant precisely because they are hard to falsify.

That tension does not make needs-based allocation unserious. It makes it vulnerable to suspicion. A government can genuinely worry about coastal unemployment and food access while also recognising that the language of need strengthens its hand at the negotiating table. Rivals, in turn, may hear every appeal to vulnerability as a disguised demand for a larger quota. This is one reason allocation debates persist even when conservation advice is widely accepted. The parties are not only dividing a harvest. They are deciding which story about fairness deserves operational priority. Historical discipline, present geography, and social need each produce credible claims. They also create different winners.

The existence of multiple plausible principles explains why fisheries bargaining so often becomes structurally unstable. If only one principle were regarded as legitimate, disputes would centre on evidence: what were the catch histories, where exactly was the stock distributed, how severe was the social dependence? But fisheries governance rarely enjoys that simplicity. Parties disagree over evidence and over the principle through which evidence should be interpreted. A state defending historical catch may downplay recent redistribution. A state favouring zonal attachment may emphasise current presence while minimising long-term investment by others. A state invoking need may ask the system to compensate for disadvantages not visible in biomass maps at all. Once principles multiply, no calculation feels neutral. The formula itself becomes a political battlefield.

A second named example sharpens this point: the Joint Norwegian-Russian Fisheries Commission, established in 1976, has often managed shared Barents Sea stocks with greater durability than many other northern arrangements. Its relative stability does not mean the allocation problem disappears. It means the parties have built procedures strong enough to keep principle disputes within a workable frame. The system has depended on repeated negotiation, scientific co-ordination, and a degree of mutual acceptance that neither pure historical entitlement nor pure zonal attachment will settle every question by itself. That is a useful counterweight to more explosive cases. It shows that mixed principles can function where institutions are credible and parties value continuity enough to accept partial dissatisfaction. Yet the limitation is obvious. Such arrangements are fragile under broader geopolitical strain, and their durability cannot simply be transplanted elsewhere by formula.

What matters most in practice is not only which principle is chosen, but what incentives it creates on the water. Historical catch can encourage racing and over-reporting if fleets suspect that today's landings will justify tomorrow's share. Zonal attachment can encourage boundary-hugging and opportunistic effort in newly favoured waters because actual catching becomes a way to reinforce a territorial narrative. Needs-based claims can, at times, intensify domestic political pressure to avoid compromise, since a government that retreats may appear to have betrayed communities presented as especially vulnerable. The visible behaviours are familiar: sudden effort concentration, resistance to seasonal restraint, strategic logbook reporting, exaggerated insistence on local dependence, or refusal to accept interim arrangements for fear that temporary concessions will harden into precedent. Allocation rules are not passive reflections of justice. They are machines for producing conduct.

This is why monitoring becomes entangled with principle. A debate over allocation can never remain entirely philosophical because each principle makes accurate reporting more or less valuable to different actors. If future shares may depend on demonstrated catch, data acquires political weight beyond compliance. If zonal presence matters, vessel tracks and haul locations become evidence in a sovereignty-inflected dispute. If need is part of the argument, landing patterns, local employment, and domestic supply assumptions become politically charged records rather than neutral statistics. Regulators then face a compounded task. They must monitor not only for conservation breaches, but for strategic behaviour directed at the next allocation fight. The argument over rights begins to shape the incentives of reporting itself.

A further trade-off appears between stability and adaptability. Historical catch can make a system more predictable, which encourages investment and reduces the frequency of destabilising renegotiation. It can also make the system sluggish and unjust under ecological change. Zonal attachment can help a bargain adapt more quickly to movement, but it can make every shift in distribution a fresh site of contest. Needs-based adjustments can improve legitimacy and social resilience, but they may weaken the sense that access rests on transparent, consistently measurable criteria. None of these problems is accidental. They arise because allocation must perform several jobs at once. It must conserve political order, reflect ecological reality, sustain investment, and remain publicly defensible. No single principle completes all four tasks gracefully.

The temptation, therefore, is to look for hybrid formulas, and often that instinct is wise. Fisheries regimes sometimes blend historical catch with spatial indicators, or temper market logic with social conditions, or stabilise concessions through linked package deals beyond quota numbers themselves. These mixed approaches can reduce the brutality of pure principle. They can also multiply ambiguity. Once several logics are combined, every party can accuse every other of cherry-picking the part of the formula that suits them most. Hybrid design is not an escape from politics. It is politics admitting that no simple rule will command assent on its own. That admission can be mature. It can also be exhausting.

Still, one should resist the conclusion that allocation is hopeless because it is contested. The deeper lesson is more exacting. Allocation persists as the central political problem of fisheries management precisely because it lies at the junction of ecology, law, investment, and identity. Fish move. Institutions remember. Communities depend. Governments calculate. Total catches may remain biologically sustainable while every surrounding question becomes sharper: who catches, who lands, who leases, who speaks for the fishery, who absorbs ecological surprise, who bears the cost of waiting. A management regime that settles the science but evades these distributive issues has not solved the fishery. It has merely postponed the moment when the distributive struggle will force its way back to the surface.

The quiet motif here is the ledger. It records past catches, current distributions, and claims of need, but it never decides by itself which column should carry the most moral weight. That choice belongs to institutions and bargaining power. It also belongs, uncomfortably, to the stories states tell about what they have earned and what they can no longer afford to lose. Once fish are translated into quota shares, the ledger stops being a technical instrument and becomes a constitutional document for the fishery.

The unresolved question follows naturally and never stays quiet for long. When a stock remains sustainably fished but its value must still be divided among actors with different histories, geographies, and vulnerabilities, which principle deserves precedence: the one that protects reliance, the one that tracks present distribution, or the one that responds to need even when it cannot be measured with the same apparent precision as catch and location?

Zonal attachment in a warming Arctic

In 2010, Iceland set a unilateral mackerel quota after the stock had become markedly more available in Icelandic waters during the summer feeding season. The fish were there. That much was hard to deny. Yet the dispute that followed was not settled by pointing at the water. Other coastal states could accept the biological shift and still reject the political conclusion Iceland drew from it. The argument turned on a deceptively simple phrase: where the fish are. Simple phrases are dangerous.

Zonal attachment has obvious appeal because it promises to align fisheries rights with present ecological reality. If fish spend more time in one state's exclusive economic zone, why should that state not receive a larger share of the quota? The principle sounds empirical rather than historical, fair rather than inherited, adaptive rather than complacent. In an Arctic and sub-Arctic shaped by moving stocks, that promise is powerful. It offers a way to revise old bargains without pretending the sea has stood still. It lets law look outward, towards current distribution, instead of backward, towards older catch records and older maps of abundance. For governments that can now point to fish in their waters, zonal attachment feels like common sense formalised.

The trouble begins as soon as one asks a harder question. Which fish, where, and when? A stock is not a pile of property resting quietly inside a line. It moves by season, depth, temperature, prey, spawning behaviour, and age class. It may feed in one zone, overwinter in another, spawn in a third, and be commercially accessible at different moments to different fleets. A survey can measure biomass in one season and produce one political claim. Another survey, timed differently or aimed at another life stage, can yield a different one. The principle of zonal attachment is clear enough in the abstract. The facts to which it attaches are not. The map is still. The stock is not.

That distinction explains why zonal attachment can stabilise bargaining in some fisheries and inflame it in others. Where parties broadly agree on how to measure distribution, over what period, and with what weight attached to seasonal presence, the principle can make renegotiation more credible. It allows a management system to respond to ecological change without tearing itself apart. A state that gains access sees a path to recognition. A state that loses some access sees a rule, however unwelcome, rather than a naked political defeat. The principle works best when its evidentiary basis is accepted as legitimate before the numbers begin to bite.

The northeast Atlantic mackerel disputes after 2010 are the canonical reminder that this acceptance cannot be assumed. Iceland and the Faroe Islands argued that the stock's increased presence in their waters justified substantially larger shares. Their counterparts, especially the European

Union and Norway, did not simply deny movement. They disputed the significance of the movement for allocation. Was summer feeding presence enough to warrant a major redistribution? Should a stock's location in one part of the year outweigh long-standing historical catches or older distributions across the rest of the annual cycle? Could a few years of altered availability carry the same legal and political weight as decades of established participation? Zonal attachment did not fail because the fish were absent. It failed because presence alone could not answer what kind of presence mattered.

The concept becomes still more slippery in northern demersal fisheries. Consider cod in the Barents Sea and adjacent waters. Cod distribution responds to temperature, prey, spawning conditions, and other ecological pressures, but not in the clean manner implied by legal language that imagines a stock occupying a zone like a tenant in a flat. Dense concentrations may appear in one area while spawning grounds remain elsewhere. Juveniles may use one habitat, larger fish another. A stock can shift its commercially valuable centre of gravity without undergoing a simple geographic migration visible on a single annual map. For one fleet, "where the fish are" means where catch rates justify steaming and gear costs. For another, it means where the fish are measured in surveys. For a government, it means where a legal argument can be anchored. These are not identical locations.

This is why zonal attachment is never a purely ecological principle. It is also always a principle of measurement. The show is familiar: a coastal state points to survey data, vessel records, or local abundance; the claim is called spatial fairness; the explanation is that quotas should track distribution; the implication is that every contest over timing, depth, and stock components becomes a contest over entitlement. Once quota numbers depend on which spatial measurement is chosen, the methodology is no longer a technical preface to bargaining. It is bargaining by other means.

A second named example reveals the same dynamic from a different angle. The Agreement to Prevent Unregulated High Seas Fisheries in the Central Arctic Ocean, signed in Ilulissat in October 2018, was precautionary, in part, because the parties recognised how unstable future allocation arguments could become if commercially important stocks emerged in areas with limited agreed-upon evidence about distribution, productivity, and accessibility. The agreement did not solve zonal attachment in advance. It postponed the rush to convert uncertain future presence into immediate claims. That restraint was telling. It acknowledged that in the absence of trusted measurement, tying rights too quickly to location could create conflict faster than governance capacity. Even here, where the fishery had not yet matured, zonal attachment hovered in the background as both promise and warning.

The attraction of the principle remains real because historical catch alone can become morally and politically threadbare under ecological change. If a stock now spends meaningful time in waters long marginal to the fishery, and if local fleets can catch it lawfully and efficiently, insisting that older participants retain nearly all the value begins to look like fossilised privilege. Zonal attachment offers a language of adaptation. It tells gaining states that the legal order can acknowledge what the ecosystem is doing. It tells losing states that revision, however painful, need not be framed as arbitrary confiscation. In this sense the principle can stabilise bargaining precisely because it gives ecological change a recognised route into law.

And yet the same principle can intensify conflict because it rewards proof of presence at politically convenient moments. A state expecting future claims to be strengthened by current observations has every incentive to emphasise the season, area, or survey frame that flatters its case. Fleets may race to establish practical use of newly favoured waters. Regulators may press for measurement regimes that support their preferred interpretation of distribution. Scientists may find their methodological caution drawn into arguments that no longer feel merely scientific. The visible behaviours are telling: sudden official emphasis on local survey results, intense debate over seasonal weighting, hardening insistence that one year's observations must not be dismissed, or, conversely, that they must not be trusted too soon. Zonal attachment invites empirical argument. It also raises the stakes of every empirical dispute.

The difficulty becomes sharper because ecological variability does not respect legal simplicity. Law likes stable units. It needs boundaries, categories, and decision rules that can be defended in writing and applied by institutions. Ecology, especially in warming northern seas, often offers gradients, pulses, overlaps, and partial redistributions. A stock may expand into new waters without abandoning its older core. It may move northward in one life stage and remain conservative in another. It may become more accessible in a zone not because more biomass resides there annually, but because changing conditions produce short, intense feeding periods that local fleets can exploit. The legal temptation is to compress this fluidity into a single governing fact. The political consequence is that whoever dislikes the chosen compromise will call the resulting bargain illegitimate.

A credible limitation must be admitted. Not every fishery requires zonal attachment to be dominant, and not every ecological shift should trigger immediate recalculation. Some stocks fluctuate around older patterns in ways that would make constant reallocation destabilising and opportunistic. Institutions need continuity as well as responsiveness. If every temporary pulse of abundance in one zone produced a new quota demand, co-operative management would become little more than permanent renegotiation under seasonal panic. This is the strongest caution against an overenthusiastic use of zonal attachment. Ecological movement must matter, but not every movement should carry the same legal consequence.

That limitation does not weaken the case for zonal attachment so much as refine it. The serious question is not whether location matters. It plainly does. The serious question is how location should matter. Over what time horizon should distribution be assessed? Which seasons deserve regulatory priority? Should feeding presence count as heavily as spawning distribution? How should depth and age structure be handled when a stock's legal unit fails to capture its ecological diversity? Can temporary pulses be distinguished from durable shifts without simply handing incumbents a permanent veto? These are not side issues. They determine whether zonal attachment functions as an adaptive principle or as a recurring invitation to dispute.

Institutional trust therefore becomes central. Where parties trust one another's surveys, accept joint scientific work, and believe that revision can occur without immediate political humiliation, zonal attachment can serve as a relatively stabilising rule. It makes adaptation legible and gives ecological change a recognised place in bargaining. Where trust is weak, the same principle becomes incendiary. Every survey is accused of bias. Every seasonal concentration is treated either as decisive proof or as opportunistic overreading. Every delay in revision becomes evidence of bad faith. Under those conditions, "where the fish are" no longer describes a shared object of measurement. It becomes a diplomatic slogan.

One should also notice how zonal attachment changes fleet incentives. If current distribution influences future share, fleets and governments gain reasons to document presence aggressively, maximise local landings, and resist interim arrangements that might understate newly favoured access. This can produce useful investment in science and monitoring, but it can also encourage boundary fishing, selective reporting, or pressure to fish hard in areas where ecological presence is politically valuable even if commercial conditions are marginal. The principle therefore reaches beyond negotiation rooms. It shapes conduct at sea and in port. Allocation logics always do.

The deeper issue is constitutional. Zonal attachment asks whether fisheries rights should track moving ecology closely enough to preserve legitimacy under change, or whether rights must remain partly insulated from movement in order to preserve continuity and reduce opportunism. Neither answer is absurd. A system too rigid to acknowledge redistribution loses moral authority and invites unilateral action. A system too sensitive to every redistribution risks becoming ungovernable. The legal simplicity everyone wants - fish are here, so quota should follow - collapses under the weight of seasonal migration, mixed stock structures, and contested evidence. But abandoning the principle altogether simply restores the older inequities that gave it force in the first place.

The quiet motif throughout is the line. On the chart it looks clean. In the water it means much less than governments wish. Fish cross it by season, by depth, by prey field, by ordinary biological indifference. Yet the line still matters because law, entitlement, and public expectation continue to gather around it. Zonal attachment is the attempt to make that line more ecologically honest without dissolving its legal usefulness. It is an elegant ambition. Elegance does not make it easy.

The next dispute over a moving northern stock will almost certainly contain this argument at its centre, whether named or not. When distributions vary with season and depth, and when the fish are simultaneously in one state's waters, another's history, and a third's market chain, what should count as the legally relevant location: the place where biomass is measured, the place where the fish are caught, or the place where a bargain can still survive being adjusted to fit an ecosystem that refuses to stay still?

Transferability, leasing, and consolidation

In 1990, New Zealand introduced its Quota Management System in a form that made transferable rights central to the organisation of commercial fishing. The practical result was immediate and far-reaching. Quota ceased to be only an annual administrative permission and began to behave more like an asset: something that could be held, sold, leased, financed, aggregated, and used to restructure the economics of the fleet. For policy designers, this was an achievement of order. For many fishing communities, it marked the beginning of a quieter unease. The paper moved faster than the harbour.

That is the essential tension of transferable quota. It promises to solve a genuine problem. Fisheries governed by rigid, non-transferable rights often trap effort in the wrong vessels, reward haste over planning, and prevent quota from moving to operators best able to fish efficiently under changing ecological and market conditions. Transferability answers with a clean proposition: let rights move to where they are used most productively. Leasing deepens that flexibility by allowing temporary reallocation without permanent sale. Consolidation follows almost naturally, because rights tend to gather where capital is patient, management is sophisticated, and economies of scale can be exploited. None of this is imaginary. It can reduce waste, lower unit costs, smooth seasonal operations, and make fleets more adaptable when stocks shift or markets tighten.

The economic argument is persuasive because it describes visible improvements. A vessel owner unable to catch a full entitlement can lease quota out rather than leave it idle. A more efficient operator can lease or buy additional quota rather than race other vessels in a short season. Processors gain predictability when rights are held by firms able to plan landings rather than chase frantic volume. Banks prefer fish businesses whose access rights look stable and tradable. Ministries like systems that convert political allocation into something more self-adjusting. The name for this is efficiency, but it is worth being precise. Efficiency here means fewer vessels chasing the same catch, more deliberate trip planning, more predictable landing schedules, and a stronger capacity to match fishing effort to both stock condition and market opportunity. These are real gains. So are the political costs they can conceal.

The first of those costs lies in the separation of ownership from place. Once quota becomes transferable, the legal holder of the right and the community that historically depended on the fishery no longer have to coincide. Rights can drift away from the harbour even while the fish remain culturally central to it. Leasing can soften the break for a time, but it may also intensify it by creating a class of operators who work the fishery without owning durable access and a class of owners who draw income from the quota without fishing it themselves. The system remains lawful, even elegant on paper. Yet to a coastal town watching control move into distant companies, funds, or urban offices, elegance is not the word that suggests itself.

Iceland offers one of the clearest illustrations of both the strength and the strain of this model. Since the late twentieth century, the Icelandic individual transferable quota system has often been praised for improving biological and economic performance in key commercial fisheries. That praise is not empty. The system helped align capacity with available catch and created a disciplined rights structure in a country whose economy and political life are unusually shaped by the sea. But Iceland has also lived the other side of quota transferability: strong public controversy over concentration, over the distance between formal ownership and local dependence, and over whether the nation's fish had become, in effect, a private asset too cheaply detached from a common inheritance. The dispute was never only ideological. It was anchored in visible outcomes: who could still enter the fishery, which communities retained landings, who paid leasing costs, and who appeared to prosper from quota without sharing the burden of fishing itself.

That dispute reveals the deeper political issue. Transferability changes not only economic performance but moral perception. When quota behaves like property, arguments about fairness start to change tone. Owners speak the language of investment security, efficiency, and legitimate expectation. Communities losing control speak the language of exclusion, enclosure, and democratic deficit. Both vocabularies have force. The governance problem is not that one side is entirely right and the other deluded. It is that each side sees a different fishery. One sees a scarce resource better managed through flexible rights and disciplined investment. The other sees a social foundation being converted into a tradable abstraction. The same quota can look like prudence from a balance sheet and like dispossession from a quay.

Leasing intensifies this ambivalence because it makes the system work while also making its tensions more visible. In strict economic terms, leasing is often the lubricant that keeps quota systems from becoming too rigid. It allows new entrants, smaller operators, and seasonal fishers to access quota without the capital needed for outright purchase. It helps rights move quickly when stock conditions or business circumstances change. It can reduce idle quota and make mixed fisheries more manageable when catch composition is uncertain. Yet leasing also creates dependency. A fisher who must lease quota year after year may remain permanently exposed to prices set elsewhere and to owners whose incentives lie in rent extraction rather than in sustaining a local fleet. The annual transaction can begin to feel less like flexibility than like tribute.

This is where consolidation becomes politically explosive. Concentration of quota in fewer hands can improve fleet rationalisation and investment discipline. It can also hollow out local autonomy. A community may still host the vessels and workers, but strategic control over access migrates upward and outward. Once this happens, the fishery becomes more vulnerable to decisions made for portfolio reasons rather than place-based ones: a company chooses one landing point over another, shifts processing, idles a vessel, or uses quota ownership to bargain from a position the local crew or municipality cannot match. Consolidation does not merely change market structure. It changes who gets to decide whether the harbour remains alive.

The paper moved faster than the harbour. That quiet fact returns because it captures the asymmetry at the heart of quota markets. Legal transfer is quick. Social adaptation is slow. Rights can be sold in a season; community capacity takes generations to build and, once lost, is hard to recover. Policy designers often treat this as an unfortunate but secondary issue, something to be managed with side rules once the main architecture is in place. That understates the matter. In fisheries, especially in northern and peripheral regions, community legitimacy is not a decorative concern. It shapes compliance, political trust, and the willingness of people to accept scientific restraint or procedural compromise later on. A system felt to be fair can survive bad years. A system felt to have sold the fish out from under the town will struggle to command sacrifice.

This is why transferability cannot be assessed only through economics. It also has to be judged through access, entry, and political voice. Who can still become a fisher once quota prices rise? Who can remain one without perpetual dependence on leasing markets? Which communities retain meaningful control over their relationship to local stocks? Which species continue to support local food systems rather than disappearing entirely into export channels organised elsewhere? These are not sentimental questions. They affect how people understand food security itself. A government may point to strong export performance and healthy aggregate revenues while a coastal settlement experiences diminished local landing, higher dependence on outside owners, and weaker access to fish as food. Under those conditions, market success does not settle legitimacy. It may sharpen the argument against it.

A useful counter-example helps to keep the analysis honest. Not every transferable quota regime inevitably destroys community connection, and not every form of consolidation is socially corrosive. Some systems use owner-operator rules, regional landing obligations, community quotas, or transfer restrictions to preserve local presence while still allowing some market flexibility. Alaska's community development quota arrangements, though distinct from pure individual transferable quota systems, are often cited because they show that rights can be structured to channel benefits towards designated communities rather than allowing unrestricted drift to the highest bidder. Such arrangements are not simple, nor are they free of political contest. But they demonstrate that the choice is not only between rigid non-transferability and socially indifferent markets. Institutional design matters.

Still, design is always working against pressure. Once quota acquires capital value, powerful incentives emerge to loosen restraints on transfer, widen leasing options, and treat fish rights as financial instruments whose best use is determined by market exchange rather than public purpose. Regulators who introduce social safeguards therefore face continuous counter-pressure from firms and owners who can plausibly argue that excessive restrictions destroy the very efficiency gains the system was meant to produce. They are often right, at least partly. A quota regime burdened with too many immobility rules can trap rights in unproductive use, encourage hidden side payments, or preserve an uneconomic fleet under the banner of community protection. That is the limitation one must take seriously. Social acceptability purchased at the price of chronic inefficiency may not endure either.

The harder task is to see that efficiency and legitimacy are not opposing planets. They interact. A socially brittle quota system may eventually undermine the economic order it was designed to create because public backlash invites re-regulation, litigation, or political deals that erode confidence in the rights structure itself. The opposite is also true. A system that preserves community participation, transparent entry, and a visible local stake in the fishery may generate stronger long-term compliance and less political volatility, even if it sacrifices some short-run allocative neatness. Markets work inside institutions, not above them. When a quota regime loses its social licence, its formal elegance becomes a liability.

The compliance implications are especially important. Fishers who feel permanently squeezed by leasing costs or excluded from meaningful ownership may see regulatory demands through a harsher lens. Reporting obligations, monitoring rules, and conservation measures then cease to look like shared discipline and start to feel like burdens imposed on labour while rents accrue elsewhere. This does not excuse misreporting or evasive behaviour. It explains why legitimacy belongs inside the enforcement picture. A fisher working on leased quota for a distant owner has a different relationship to rule-following from one who experiences the fishery as a durable local institution. The difference may not always surface in headline disputes, but it accumulates in attitudes to regulation, co-management, and political compromise.

Food security threads through this as well, though often in muted form. Quota markets are usually justified through their effect on commercial efficiency, not through their effect on who can still access fish locally. Yet in many northern and coastal regions, the question of whether fish are landed, processed, sold, or retained locally matters well beyond commercial accounting. When transferable systems encourage concentration and remote ownership, they can weaken the practical link between nearby fish stocks and nearby food systems. A state may continue to harvest abundantly while local residents feel that the resource no longer feeds them in any meaningful political sense. That perception can become explosive when policymakers continue to speak of national fisheries strength as though aggregate value were the same thing as broad-based benefit.

The legal form of quota transfer also changes bargaining within the state. Once rights are held by firms able to mobilise finance, lawyers, and lobbying power, reform becomes harder. Governments contemplating reallocation, community safeguards, or anti-concentration measures face not just policy arguments but property expectations. Owners claim stability. Regions claim justice. Ministries try to balance both while knowing that any major adjustment will be described by one side as a necessary correction and by the other as confiscation. The initial allocation decision thus keeps governing long after it appears settled, because transferability deepens the stakes of any later revision. Quota becomes not merely a permission to fish, but an entrenched position in the political economy of the coast.

What emerges from all this is not a simple verdict for or against quota markets. Transferability, leasing, and consolidation can produce real economic gains and a more adaptable fleet. They can also produce a fishery whose legal order outruns its social foundations. The central question is therefore constitutional rather than technical. How much flexibility can a system permit before communities cease to recognise the fishery as something they meaningfully share? How much protection can it impose before the system becomes too rigid to adapt to moving stocks, changing markets, and shifting seasons? Those are not secondary design details. They determine whether a quota regime remains governable.

The paper still moves faster than the harbour. It always will. The real test is whether institutions can slow that movement enough to keep rights connected to the places, people, and political bargains that make fisheries management more than a market in abstract claims. When the next reform promises cleaner incentives, better investment, and smoother adjustment through transferable quota, what should weigh more heavily: the efficiency gained by letting rights flow to capital and scale, or the legitimacy lost when the harbour begins to feel that it no longer owns even a meaningful share of its own future?

Racing, high-grading, and discards

On 1 January 2019, the European Union's landing obligation applied fully to catches subject to catch limits in fisheries such as the mixed demersal fisheries of the North Sea. The formal outcome sounded simple enough: fish that once might have been discarded at sea now had to be landed and counted. Yet the significance of that rule lay in what it revealed rather than what it solved. It exposed a truth long familiar to fishers and regulators alike. Waste at sea is rarely caused by carelessness alone. More often it is produced by the design of the fishery itself - by quotas that do not match the species mix in the net, by incentives that reward speed over selectivity, and by the fear that what is not claimed now may not be claimable later. The fish are sorted twice.

That second sorting matters. First comes the sea's own mixture: species, sizes, and qualities brought together by habitat, season, and gear. Then comes the human sorting driven by law, price, quota, and anticipation. A fisher who catches more than can legally be landed, or catches fish of lower value that may crowd out higher-value fish later in the trip, faces a temptation that management systems often pretend away. Keep the better fish. Shed the rest. Record selectively. Fish harder now because future access may depend on recent performance. The official categories call these discards, high-grading, and racing. The deeper issue is political economy. Poorly aligned rules turn the deck into a site of bargaining with the future.

Racing appears first because it is the most visible. When a fishery is governed by short seasons, competitive openings, or allocation systems that reward demonstrated catch, operators gain from moving quickly, even if that speed worsens selectivity and safety. The show is familiar: vessels rushing to grounds as soon as conditions open, gear deployed under pressure, less time to avoid juvenile fish or mixed catches, more tolerance for borderline choices because delay carries immediate cost. The name for this is the race to fish. The explanation is straightforward. If one vessel hesitates, another may fill the market, the season, or the political record. The implication is corrosive. A management system intended to control total catch can end up encouraging the very waste and distortion it publicly condemns.

High-grading follows from a different but related incentive. A fisher holding limited quota, storage space, or landing opportunity may prefer to retain only the most valuable fish - larger individuals, cleaner target species, or catches in the best market condition - while discarding fish that are legal but less desirable. High-grading is not random waste. It is selective optimisation under constraint. The problem becomes acute in mixed fisheries where species composition varies from haul to haul and where one weakly held quota stock can choke the entire trip. A vessel may then keep fish that preserve commercial value and quietly discard those that consume quota without preserving revenue. The visible behaviour is sorting by value rather than by ecological or legal logic. The implication is that reported landings stop describing actual mortality.

The North Sea mixed demersal fisheries offer a clear example because they combine quota complexity with unavoidable species overlap. A trawler targeting haddock, saithe, or whiting may also catch cod, juvenile fish, or other species for which the operator holds insufficient quota or faces stronger regulatory pressure. Before the landing obligation, discarding provided a brutally simple release valve. After 2019, that valve was meant to close. Yet closing it did not abolish the underlying incentive. It shifted the struggle into monitoring, exemptions, survivability arguments, choke-species debates, and the practical difficulty of proving what happened in each haul. The rule changed. The deck did not become morally transparent.

A second named example shows the same logic under a different regulatory culture. In the Bering Sea pollock fishery, especially once Chinook salmon bycatch became a major management issue and caps were tightened after the North Pacific Fishery Management Council acted in 2010, vessels faced a familiar tension: continue targeting a lawful and valuable stock while avoiding incidental capture that could trigger constraint for the wider fishery. This did not produce high-grading in the classic size-selective sense alone, but it did expose the same governing mechanism. When one stock or species mix threatens to curtail access to another, reporting, sorting, and discard practices become politically charged. A haul ceases to be only a catch. It becomes evidence. Under those conditions, monitoring systems are never neutral administrative tools. They are the battleground on which competing versions of the fishery are tested.

That is why allocation design matters so much. If future quota shares depend significantly on past landings, fishers receive a signal that can be perfectly rational and environmentally destructive at the same time: catch and declare as much strategically useful fish as possible now, because today's record may decide tomorrow's entitlement. If stock distributions are shifting, the pressure intensifies. A fleet encountering a stock more frequently in its local waters may fear that restraint will only weaken its later claim to a larger share. The name for this is strategic effort under positional uncertainty. The explanation is political rather than moral. Fishers respond to the possibility that present behaviour becomes future precedent. The implication is severe. Management creates waste not only through bad rules, but through rules that make proof of participation too valuable.

Discards rise most sharply where the legal categories no longer match what the fishery is actually catching. A vessel fishes for one stock, but the size distribution changes. Smaller fish enter the grounds. A protected stock overlaps with the target. Market conditions favour one grade over another. Quota holdings fit last year's pattern, not this week's haul. Under those circumstances, discarding is often treated publicly as straightforward cheating, and sometimes it is. But it is more analytically useful to see it as a symptom of misalignment. The show is the same net bringing up a commercially mixed catch. The name is regulatory mismatch. The explanation is that fishery rules are asking operators to produce a clean outcome from an ecologically untidy reality. The implication is that enforcement becomes a fight not only over conduct, but over whether the rules themselves have become too dishonest to obey without loss.

Monitoring therefore moves to the centre. Once managers suspect that fish are being high-graded, discarded, or selectively recorded, ordinary trust in landings data collapses. More observers are demanded. Cameras are installed. Logbooks are audited against vessel tracks, hold capacity, and port records. Gear rules are tightened. Move-on provisions multiply. None of this is bureaucratic excess in the abstract. It is a response to a system in which under-reporting and selective retention can alter both stock assessments and future bargaining positions. The fish are sorted twice, and so are the data: once by those who catch them, once by those trying to infer what was hidden. Monitoring becomes central because without it, quota systems drift into fiction.

Yet monitoring is not a clean cure. Human observers can be partial, pressured, or absent from key trips. Cameras produce enormous amounts of footage that require staff, time, and agreed evidentiary rules to interpret. Electronic records may show where a vessel went but not what it threw overboard. Port inspections can verify landings, not discarded mortality already gone. The politics then intensifies. Fishers argue that monitoring regimes are punitive, impractical, or blind to ecological reality. Regulators argue that without strong verification, discards and high-grading will remain a permanent shadow fishery. Both claims can be true at once. That is why monitoring becomes a battleground rather than a settlement. It embodies a prior dispute over whether the system is meant to trust or to prove.

A credible limitation must be admitted. Not all discarding reflects cynical profit-seeking, and not all racing comes from quota design alone. Bad weather, gear selectivity limits, biological variability, and genuine uncertainty about species composition can push vessels into situations no management system handles elegantly. Some fisheries have reduced racing and waste through rights-based systems that give operators more time and clearer incentives to fish carefully. Individual transferable quotas, cooperative arrangements, and better aligned bycatch provisions can reduce the pressure to rush and may lower discard incentives. That is a real counter-example to any crude claim that quota systems inevitably create waste. The difficulty is that such regimes often solve one incentive problem while creating others - concentration, dependence on leasing, and disputes over entry and legitimacy among them.

The more exact conclusion is that waste follows from badly matched incentives, not from a single institutional form. Competitive openings can encourage racing. Rigid quota portfolios can encourage high-grading. Future allocation rules tied too closely to present landings can encourage strategic reporting. Species-mix volatility can make legal compliance feel economically irrational. Under any of these conditions, the most visible waste - dead fish discarded, landings misreported, valuable catch selected against weaker stock rules - is merely the surface expression of a deeper design failure. Fisheries governance then oscillates between moral condemnation and technical repair without fully admitting that the rules themselves may be producing the behaviour they later criminalise.

This is especially damaging because discards and high-grading corrode the information on which the whole system depends. Stock assessments assume that removals are known within tolerable bounds. Quota negotiations assume that catch histories mean something. Enforcement assumes that declared landings broadly reflect actual fishing. Once waste and under-reporting spread, those assumptions weaken together. A discarded fish is not only an ecological loss. It is a statistical disappearance. A selectively recorded haul does not only benefit one vessel but it distorts future science and future politics. The invisible catch enters the next round of management as false calm.

The public response to this often swings between two unsatisfactory poles. One blames fishers alone, as though any sufficiently virtuous operator could comply perfectly inside a system that penalises clean behaviour. The other blames the system so completely that personal responsibility disappears. Neither is adequate. Fishers make choices, sometimes poor and sometimes deliberate. Institutions make choices too, especially when they tie access, value, and future claims to behaviours that can be gamed more easily than they can be verified. The serious task is to connect those choices without sentimentality. One must ask not simply who discarded, but what made discarding, racing, or high-grading the rational temptation in the first place.

The quiet motif is the sorting table. It is where legal categories, commercial value, ecological mess, and political incentive all meet. A regulation may insist that every fish be counted honestly and every haul recorded faithfully. The table forces a faster judgment. Which fish fit the quota? Which fish damage the trip? Which fish matter to next year's claim? Which fish will likely never be seen by anyone ashore? Management is tested here more than in ministerial speeches, because here the incentive structure becomes tactile. The law either makes honest sorting survivable, or it quietly invites the second sort that nobody wants to admit.

That is why allocation and monitoring cannot be separated. A quota regime that creates fear of future loss will often invite short-term maximisation. A monitoring regime that cannot verify what happens during the most pressured moments will struggle to contain it. Rules written for annual stability may fail in the very trips where size composition changes or species overlap becomes commercially decisive. Under those conditions, discards rise not because the fishery lacks rules, but because the rules are fighting the wrong battle.

The unresolved question is therefore not whether waste is bad. That is too easy. The harder question is what kind of system can reduce racing, high-grading, and discards without simply replacing one distortion with another: a regime tight enough to make concealment costly, yet flexible enough that fishers do not experience honest reporting as a direct threat to their income, their quota position, or their claim to a share of the fishery's future.

Side payments and package deals

In September 2010, Norway and the Russian Federation signed the Treaty concerning Maritime Delimitation and Cooperation in the Barents Sea and the Arctic Ocean. The agreement is usually remembered for drawing a line after decades of dispute. That memory is too narrow. What made the settlement durable was not the line alone, but the broader structure around it: co-operation, resource management, and the practical recognition that an allocation dispute rarely yields to arithmetic by itself. A quota number can end an argument on paper. It rarely ends the bargaining beneath it. Something else is usually being traded.

That is the quiet logic of side payments and package deals in fisheries politics. When parties cannot agree on a quota share as a matter of principle, they often move sideways rather than forward. Access to another stock, reciprocal fishing opportunities, joint ventures, scientific co-operation, port preferences, phased implementation, enforcement understandings, or research concessions can be linked to the main bargain so that each side leaves with something it can present as gain rather than capitulation. The package does not eliminate conflict. It redistributes it across several files until the overall trade becomes politically survivable. The fishery is settled, temporarily, by more than fish.

This matters because allocation arguments are frequently too sharp to be solved by single-variable reasoning. One party wants a larger share because the stock has shifted. Another wants to preserve historical participation. A third fears setting a precedent that will weaken its position elsewhere. If the negotiation is framed only as "how much quota for whom", the result is often stalemate. Linked concessions alter that geometry. Show a hard distributive problem, name the tactic as issue linkage, explain that actors can exchange value across different domains when direct compromise is too painful, and the implication follows immediately: what appears to be a quota settlement may actually rest on a bundle of side bargains that carry the real political weight.

The most obvious version is an access swap. A state that cannot win all it wants on one stock may secure access to another fishery, another zone, or another seasonal opportunity. Such swaps are not merely diplomatic ornaments. They convert a zero-sum fight over one resource into a broader calculation about total benefit. A coastal state may tolerate a smaller share of a contested pelagic stock if its vessels gain access elsewhere, or if a partner opens a port, relaxes a technical measure, or accepts a joint management arrangement that reduces future uncertainty. Access becomes currency. This is one reason fisheries negotiations so often look opaque from the outside. The real bargain is not always in the quota table.

Joint ventures serve a similar function, though with a different political tone. Instead of trading access directly between states, parties encourage or tolerate commercial arrangements that spread the benefits of a disputed stock across firms, regions, or fleets that might otherwise remain locked in distributive opposition. A state that feels short-changed in formal allocation may still gain processing work, vessel contracts, crew opportunities, or investment through a joint enterprise. The name for this is compensation through commercial integration. The explanation is practical. If a party cannot secure the fish in law, it may secure part of the value chain in business. The implication is double-edged. Package deals can lower diplomatic temperature by creating shared interests, but they can also move politically sensitive decisions into commercial arrangements that remain vulnerable to domestic suspicion.

The North Atlantic offers repeated illustrations of this broader repertoire, even where the individual bargains have differed in content and tone. The northeast Atlantic mackerel disputes after 2010 are often cited as examples of allocation failure, and rightly so. Yet even in these disputes, the pressure to consider broader packages never disappeared. Scientific co-operation, reciprocal restraint, timing of adjustments, and the possibility of linked concessions all hovered around the central arithmetic. That they failed to stabilise the dispute fully is not evidence that such tools are irrelevant. It is evidence that side payments work only when the wider package remains politically legible to the parties that must defend it. A bargain that looks balanced to negotiators may look like surrender to those waiting at home.

This is why research co-operation deserves more attention than it usually receives. Joint scientific work can function as a side payment even when no one calls it that. A state asked to accept an interim quota share below its preferred level may be more willing to do so if it gains a larger role in future survey design, shared access to data, or a recognised place in the institutions that will judge the next round of distributional change. This is not a trivial concession. In moving fisheries, influence over how the fishery is measured can shape later bargaining power. The show is quiet - committee seats, vessel time, common surveys, working groups - but the name is institutional compensation. The explanation is that participation in the knowledge structure can offset disappointment in immediate allocation. The implication is that scientific co-operation is never wholly separate from politics.

A second named example helps clarify the point. The 2018 Agreement to Prevent Unregulated High Seas Fisheries in the Central Arctic Ocean did not divide a mature commercial fishery, but it did bundle restraint with research and institutional co-operation. That structure mattered. The parties agreed not only to hold back from unregulated fishing but also to create a framework for joint scientific work and future consultation. In effect, uncertainty itself was packaged. No one could claim a settled distribution of benefits, so the bargain linked present restraint to future voice. That was not a side payment in the narrow cash sense, but it performed the same diplomatic function. It gave parties something to take home besides self-denial.

The central lesson is that fisheries bargains are often stabilised by trading across time as much as across issues. A party may accept less now in exchange for review later, joint monitoring, phased implementation, or a mechanism that keeps future claims alive without forcing immediate confrontation. Delay, when structured well, can itself be a concession. It allows governments to present an arrangement as provisional rather than as a final surrender of principle. This can buy diplomatic peace. It can also store trouble for later if the underlying distributive question remains untouched and the review clause becomes merely a delayed battleground. Package deals reduce pressure partly by spreading it across years.

That temporal feature explains why such bargains are often politically fragile at home. Domestic audiences tend to value visible quota shares over procedural concessions, future review clauses, or technical co-operation. A minister may know that access to shared research, to port services, or to a separate fishery materially improves the national position. A parliament or fishing region may see only that "our" share of the disputed stock appears lower than it should be. The package then unravels in public translation. A linked bargain that made strategic sense in negotiation becomes hard to defend because the non-quota elements are harder to symbolise. Fish are visible. Committee seats are not.

National fish complicate this still further. Once a species comes to stand for sovereignty, livelihood, or regional dignity, issue linkage becomes more difficult because concessions in that fishery are no longer judged only by material balance. They are judged by symbolic meaning. A government that offers or accepts a compensating package may be accused of trading away a national resource for side benefits that look technocratic, obscure, or insufficiently patriotic. The same linked bargain that a negotiator experiences as intelligent statecraft can be framed by opponents as a backroom swap. This is why package deals, though often sensible, are also vulnerable. They require a political class capable of defending complexity in public. Fisheries politics does not always reward that skill.

And yet side payments can be indispensable precisely because they acknowledge the plural nature of fisheries value. A quota is not the only thing at stake. So are landings, processing work, science, enforcement routines, access to adjacent stocks, investment flows, and the tone of wider relations. A state may accept a smaller formal share if its firms receive privileged commercial opportunities or if joint enforcement reduces future uncertainty. A coastal region may be mollified by guaranteed landing obligations or local processing commitments attached to a wider settlement. These are not evasions of the real issue. They are often the only way to make the real issue governable. The fishery exists inside an economy and a diplomatic relationship, not outside both.

A reasonable limitation must be admitted. Side payments and package deals can obscure accountability as easily as they can facilitate compromise. When too many elements are bundled together, it becomes harder to tell whether the underlying allocation is defensible or merely hidden behind a compensating haze. Commercial beneficiaries may not be the same people as those who bear the cost of a smaller quota. A research concession may sound valuable but fail to satisfy communities facing immediate landing losses. Joint ventures can distribute gains unevenly and may deepen dependence on larger or better-capitalised partners. A package that looks balanced at state level may be sharply imbalanced at regional or class level. This is the danger of diplomatic elegance. It can smooth over the very domestic fractures that later reopen the dispute.

The use of package deals also changes the conduct of negotiation itself. Parties begin to bargain not just over numbers but over sequencing, confidentiality, and signalling. A state may hold firm publicly on quota while privately searching for off-table concessions it can accept without losing face. Another may publicise only the quota element of a settlement while downplaying the compensating benefits it has also secured. The formal record then tells only part of the story. For analysts and citizens alike, the practical lesson is to ask what moved alongside the quota. Was access exchanged elsewhere? Were enforcement expectations softened?

Did research co-operation expand? Were private commercial channels opened in ways that redistribute the meaning of the agreement? The visible number is seldom the whole transaction.

This is where the broader diplomatic repertoire comes into view. Fisheries negotiations borrow tools familiar from other forms of statecraft: linkage, sequencing, issue bundling, ambiguity, and reciprocity. They are not exceptional simply because fish are biological resources. They are part of a larger world in which states manage distributive conflict by enlarging the field of exchange. The difference is that fisheries bargains are rooted in moving ecology and in domestic communities that often experience the resource personally. That makes side payments both more necessary and more delicate. An oil concession, a tariff adjustment, and a fisheries access swap may all be linked in negotiations elsewhere, but fish are unusually likely to carry local memory and national symbolism with them.

The quiet motif here is the second envelope. The first envelope contains the quota figure everyone sees. The second contains the access swap, the research role, the phased review, the joint venture, the landing understanding, the enforcement courtesy, the side arrangement that makes the visible number bearable. Serious fisheries analysis has to open both. Otherwise it mistakes the public arithmetic for the whole bargain and misses how stability is actually purchased. Sometimes the second envelope saves the deal. Sometimes it stores the resentment that will later undo it.

That is the unresolved test. When a quota bargain cannot be made acceptable on its own terms, should governments widen the negotiation until a package becomes possible, knowing that linked concessions may produce a wiser settlement but a shakier one in domestic politics? Or is a clean and narrower failure preferable to a broader diplomatic success that no minister, fleet, or coastal community can safely bring home once the second envelope is opened?

Domestic politics of "national fish"

In August 2013, the European Union imposed trade measures against the Faroe Islands during the herring and mackerel disputes that had already strained relations across the north-east Atlantic. The legal and commercial issues were real enough. But what gave the episode its force was not law alone. In Torshavn, fish were not merely quota units under contention. They were work, pride, continuity, and proof that a small polity could defend its interests against larger powers. Once that meaning took hold, compromise became harder to stage. A fish can enter parliament before it ever reaches market.

That is the central problem of the national fish. Certain species acquire a political charge far beyond their biological or even economic profile. They come to stand for sovereignty, regional dignity, coastal continuity, or the idea that a nation still commands something tangible in a world of moving markets and distant rules. This transformation does not happen by accident. It is produced through repeated argument: ministers invoking the species in speeches, regional leaders tying local livelihoods to its defence, newspapers framing quota disputes as tests of resolve, and fishing communities reading each round of negotiation as a public measure of whether the state takes them seriously. Once a species has acquired this symbolic density, the bargaining field changes. The negotiator no longer carries only a fisheries brief. The negotiator carries a flag.

The distinction matters because technocratic analyses often assume that governments enter fisheries negotiations as utility maximisers with clearly ranked preferences. In that picture, if the numbers line up and side payments are available, a rational bargain should emerge. Often it does not. The missing variable is domestic political meaning. A minister may know perfectly well that an interim compromise would conserve the stock, maintain institutional co-operation, and leave room for future revision. Yet if that compromise can be framed at home as surrendering "our fish", its formal rationality becomes politically useless. Domestic politics is not noise added to the system from outside. It is part of governance capacity itself. A state unable to carry a deal through its own public arena does not, in any practical sense, have the capacity to make that deal.

The Faroe example is revealing because the politics of fish there cannot be reduced to simple nationalism. Fisheries matter deeply to the islands' economy and social life, but the political potency of species such as mackerel and herring has come not only from dependence but from scale and memory. A small polity negotiating with much larger actors naturally reads access disputes through the lens of recognition and autonomy. The show is familiar: a stock becomes more available locally, the government presses for a larger share, external resistance follows, and public discourse turns the dispute into a test of whether local interests will be acknowledged or brushed aside. The name for this process is symbolic loading. The explanation is that a resource becomes a vessel for wider anxieties about status and control. The implication is that compromise narrows because each concession now appears to say something larger than itself.

Iceland offers a second example, especially during the mackerel disputes that intensified after 2010. Mackerel was not politically salient there only because of catch value. It became salient because the stock's altered presence in Icelandic waters seemed to dramatise a wider claim: that the ecological reality of the north-east Atlantic had shifted, and that older quota structures were denying recognition to that shift. The domestic language surrounding the dispute therefore fused fairness, sovereignty, and livelihood. This fusion mattered. Once a fishery is narrated as evidence that the nation is being treated unjustly, bargaining space contracts. A negotiator returning from talks with a technically elegant but visibly modest gain may discover that elegance is the least valued feature of the outcome.

A national fish does several kinds of work at once. It concentrates diffuse economic concerns into a single emblem. It gives peripheral regions a politically legible object around which to organise pressure. It lets opposition parties accuse the government of weakness in terms ordinary voters can grasp. It also allows governments to discipline internal dissent by presenting a complex management issue as a matter of collective defence. Fish, in other words, simplify politics while making governance harder. They turn distributive complexity into moral clarity, and moral clarity is rarely negotiable. A landing can be discussed. A symbol is harder.

This symbolic conversion is often mediated by parliament and regional lobbying. Legislatures matter not because they are always decisive in the legal sense, but because they provide the stage on which a fishery is nationalised rhetorically. Members from coastal or island constituencies tend to speak with unusual force in such debates because they can point to concrete impacts: a plant at risk, crews waiting for a season, local businesses tied to landings, younger fishers unable to enter the trade. Their interventions do more than express concern. They impose political costs on compromise by ensuring that every quota adjustment can be narrated as an injury to named places and people. The problem is not irrationality. It is representational intensity. Where a fishery touches particular districts deeply, those districts will speak loudly, and governments that ignore them do so at their peril.

Media narratives then harden the frame. Technical disputes over zonal attachment, stock assessment, or linked concessions rarely survive translation into public discourse without simplification. "Fair share" and "stolen fish" travel better than methodological caveats. Images of vessels, crews, and harbours outrun the quieter logic of package deals and provisional review clauses. Once headlines and broadcasts have established a national fish as something being taken, denied, or defended, a government that seeks compromise must fight not only a negotiating counterpart but the narrative momentum at home. This is why public communication belongs inside fisheries governance rather than outside it. Information environments alter the bargaining room. A state that cannot explain a deal before others explain it for them loses capacity in advance.

The relevant abstraction here is legitimacy, but it should be made concrete. Legitimacy is visible when a minister can return from a negotiation, present a compromise, and still expect compliance and political survival. It is visible when regional actors accept a painful adjustment because they believe the process respected them and the outcome did not simply ignore their dependence. It is visible when parliamentary criticism remains bounded rather than escalating into accusations of betrayal. The absence of legitimacy is equally visible: emergency debates, accusations that the government "sold out" a stock, amplified local protest, collapsing trust in scientific or diplomatic institutions, and demands for unilateral action even when such action may worsen the long-term position. These are not emotional aftershocks. They are measures of governance strain.

One reason national fish are so difficult to manage is that they often bind together interests that do not fully align. Export firms may want stable market access and predictable rules. Small coastal operators may want local landing guarantees and tougher resistance in negotiations. Consumers may scarcely feel the issue at all except when it is framed as a sovereignty matter. Ministries of finance, fisheries, and foreign affairs may each evaluate a proposed bargain differently. The fish becomes national precisely by compressing these differences. Publicly, everyone is defending the same species. Privately, they are defending different things through it. A government can therefore appear united while negotiating and still discover, once it returns home, that unity was only rhetorical.

A useful limitation should be admitted. Not every important fish becomes a national fish, and not every fisheries dispute is captured by symbolic politics. Some stocks remain too diffuse in public meaning, too commercially narrow, or too distant from everyday political imagination to trigger this kind of mobilisation. In other cases, institutional habits are strong enough that even valued species remain largely within technocratic management channels. The Joint Norwegian-Russian Fisheries Commission, despite all the weight attached to northern stocks in both countries, has often shown that routine procedure can contain much symbolic charge before it spills over into open political theatre. That example matters because it reminds us that nationalisation is not inevitable. It is produced under particular conditions: visibility, local dependence, media uptake, and a sense that the fishery speaks for something larger than itself.

Still, the limitation should not be overstated. Even durable institutions operate within domestic political fields that can harden abruptly when ecological change or external pressure sharpens the stakes. A species may remain largely technical until a stock shift, trade sanction, access denial, or high-profile incident turns it into a symbol overnight. Once that shift occurs, rational bargains become harder not because reason disappears, but because the definition of a rational bargain itself changes. What is rational to a negotiator may be irrational to a governing coalition that cannot survive the appearance of weakness. What is rational to a ministry concerned with long-term co-operation may be irrational to a coastal representative facing immediate electoral punishment. Rationality fragments along institutional lines.

This is why domestic politics should be treated as an element of fisheries capacity. Capacity is not only patrol reach, scientific competence, or legal authority. It includes the ability to absorb compromise without institutional panic, to communicate complex bargains credibly, and to keep parliamentary and regional opposition from collapsing every negotiation into a referendum on national honour. A government that lacks this ability may possess strong technical advice and sound diplomatic options while remaining functionally weak. It cannot convert prudent policy into durable rule. This weakness is especially acute in warming seas, where shifting distributions mean more frequent adjustment and therefore more frequent opportunities for national fish politics to reactivate.

The practical consequence is that domestic political management is never secondary to quota bargaining. Governments that neglect it often mistake a stable technical process for a stable political one. They assume that because the science is adequate and the package balanced, the outcome will survive contact with home constituencies. Often it will not. Package deals become suspect because the visible quota share appears too low. Research co-operation is dismissed as a consolation prize. Review clauses are attacked as vagueness. Market access concessions are ignored because they do not satisfy the symbolic demand to defend the species itself. The more a fish has come to stand for sovereignty or dignity, the harder it becomes to compensate for perceived loss with anything other than direct and visible gain.

The quiet motif throughout is the flag, though it need not literally fly above the harbour to shape what happens there. A species becomes national when arguments about it begin to answer questions far beyond fisheries policy: who counts, who yields, who is ignored, who governs, who still belongs to the state strongly enough to be defended. Once a fish carries that burden, every quota percentage becomes heavier. Every scientific caveat sounds political. Every diplomatic concession looks larger than its text. This is not irrational theatre laid over a fundamentally technical issue. It is the issue transformed by the way democratic polities attach meaning to material life.

That is why rational bargains fail. They fail not because negotiators are fools or publics are incapable of complexity, but because some bargains, however sensible on a spreadsheet, collide with symbols that organise political loyalty more powerfully than the spreadsheet can measure. A national fish is one of those symbols. It narrows language, shortens patience, and turns procedural compromise into an existential style of argument. Governments know this. Their counterparts know it too, and often bargain with it in mind. Domestic politics thus enters international fisheries management not as background noise but as active terrain.

The question that remains is the hardest one of all. When a species has become shorthand for sovereignty and livelihood, what should a responsible government defend most fiercely: the symbol that domestic politics demands, the compromise that long-term governance requires, or the institutional credibility that may be lost if every fisheries negotiation is forced to prove, yet again, that no national fish can ever be partially shared without looking like defeat?

Chapter 5

Institutions that Manage the Arctic

In October 2018, in Ilulissat, Greenland, ten parties signed the Agreement to Prevent Unregulated High Seas Fisheries in the Central Arctic Ocean. The striking feature of that moment was not force but restraint. The agreement did not celebrate a decisive enforcement victory. It created a structure before the full commercial contest had arrived. That choice says something basic and often forgotten about Arctic governance. Institutions matter most not when they end a crisis in public, but when they make a crisis less likely to begin.

This is easy to say and harder to honour. The Arctic attracts a style of commentary that speaks as though power begins with sovereignty and ends with patrols. But a state can possess legal rights over a zone and still struggle to convert them into orderly practice. A regional body can produce standards and still fail to compel equal compliance. A technical working group can keep difficult relationships usable for years, then suddenly lose authority when wider politics turns procedural doubt into strategic suspicion. The file on the desk matters. So does who opens it, who delays it, and who trusts what is inside.

The central tension lies here: formal authority in Arctic fisheries is often clear enough, while practical control is uneven, layered, and occasionally brittle. Coastal states regulate licensing, access, reporting, inspections, and domestic implementation within their own waters, yet those powers meet ordinary limits in budget, distance, fleet sophistication, and bureaucratic design. Beyond those waters, co-operative forums offer continuity, standards, and a language of procedure, but rarely the clean power to end disagreement by decree. Between the two sits a landscape of seams - places where one jurisdiction ends, another begins, and incentives sharpen because no single institution carries the whole burden well.

That tension matters because fisheries disputes are rarely caused by a total absence of rules. More often they grow inside crowded rule systems that do not fit together neatly enough under pressure. One agency collects data. Another issues licences. A third handles enforcement. A regional forum receives reports but cannot verify them directly. A coastal state insists it has acted lawfully while a neighbour doubts the follow-through. None of this looks dramatic in isolation. Yet fleets notice quickly when administrative friction produces usable gaps. They learn where paperwork slows, where jurisdictions overlap awkwardly, where routine co-ordination depends too heavily on goodwill, and where a contested incident may be easier to politicise because no single institutional channel commands immediate trust.

The Arctic sharpens these problems because remoteness punishes fragmentation. A missed communication between agencies is more serious where inspection opportunities are rare. A weak hand-off between science and enforcement matters more where a season is compressed and a stock is moving. A lag in regional procedure becomes more consequential where states fear that delay itself changes bargaining position. Institutions, in such conditions, do not merely regulate activity. They organise time. When they work, they slow escalation by giving actors somewhere to put disagreement besides the nearest patrol vessel or newspaper headline. When they fail, the same disagreement spills into harder forms of signalling.

This is why co-operative bodies deserve to be understood less as decorative forums than as shock absorbers. Regional fisheries management organisations, joint commissions, technical committees, and scientific working groups often look dull precisely because they are doing some of the most important political work. They standardise language. They harmonise expectations. They keep disputes procedural long enough for parties to avoid treating every setback as a test of resolve. Their weakness is equally important. Consensus can freeze adaptation. Uneven compliance can hollow out credibility. Familiarity can turn into ritual if the ecological facts move faster than the institution's appetite for revision. Still, without these bodies, many disputes would arrive harder and earlier.

The path ahead follows that layered architecture from the inside out. It begins with the powers coastal states possess in theory and the limits they encounter in practice, then moves into the regional organisations that convert disagreement into procedure without always resolving it. From there the argument passes through the high seas pockets and regulatory seams where free-riding becomes attractive and verification becomes harder, before settling on the quieter machinery of joint commissions, technical exchanges, and routine diplomatic contact that often keeps a season governable before anyone notices.

A useful caution belongs here. Institutions can absorb friction, but they can also preserve it in respectable form. A committee may delay what a fishery urgently needs. A reporting system may create the appearance of control while leaving key conduct unverified. A co-operative process may become politically sacred even after participants no longer trust its outputs. This limitation does not weaken the institutional argument. It sharpens it. Governance capacity is not measured by the existence of bodies, meetings, and rules alone. It is measured by whether those arrangements still alter behaviour under stress.

The question that follows is therefore more exacting than it first appears. When Arctic fisheries move across zones, across agencies, and across institutional levels at once, what counts as real control: the legal right to regulate, the regional forum that keeps talking, the seam where nobody can act decisively, or the ordinary routine by which officials, scientists, inspectors, and diplomats keep enough trust alive that the next argument stays in the file rather than on the deck?

Coastal state management and its limits

On 16 January 2019, the Norwegian Coast Guard boarded the Russian trawler *Melkart-5* in the fisheries protection zone around Svalbard, and the encounter soon widened into a dispute over fisheries rules, jurisdiction, and the handling of a vessel under northern conditions. The visible fact was a boarding. The more important fact was institutional. Norway did not lack legal language, patrol assets, or a fisheries administration. Yet the incident still revealed how quickly formal authority in Arctic waters can turn into a test of operational reach, evidentiary discipline, and political nerve. A state can have the rule. It may still struggle with the moment.

That is the starting point for understanding coastal state management in the Arctic. Within their exclusive economic zones, and in adjacent regimes where they assert or exercise fisheries control, coastal states possess extensive powers. They can license vessels, set conditions of access, specify gear, require reporting, impose closures, inspect catches, deny services, and pursue sanctions through administrative or criminal channels. In legal terms, this is substantial authority. In practical terms, it is only the beginning. The sea does not enforce the file. A regulation drafted in a capital becomes real only if agencies can interpret it consistently, patrols can reach the relevant waters, inspectors can document what they find, analysts can process the data, and decision-makers can carry a case through to sanction without losing nerve or procedural coherence. The signature is quick. Implementation is slow.

This gap between formal competence and lived control defines much of Arctic fisheries governance. The show is familiar: a state announces a licensing regime, a reporting obligation, or a seasonal measure. The name for the next challenge is implementation capacity. The explanation is not simply that northern waters are vast or harsh, though they are both. Capacity depends just as much on budgets, vessel availability, staffing, communications, legal process, and bureaucratic design. A coast guard may have a presence yet lack specialist fisheries expertise on every patrol. A fisheries ministry may issue sound regulations yet rely on another agency for boarding, a third for prosecution, and a fourth for data management. The implication is that the fishery is governed not by law alone but by the weakest hand-off in the institutional chain.

The Arctic magnifies every such weakness because distance punishes delay and fragmentation. A licence anomaly spotted in an office may matter only if someone at sea can act before the vessel shifts grounds, lands elsewhere, or uses poor weather as cover. A suspicious pattern in electronic reporting may need to be cross-checked against satellite tracks, observer records, and landing declarations held by different bodies that do not share data quickly. A patrol vessel may know enough to be concerned yet lack the evidentiary confidence to escalate safely. These are not dramatic failures. They are ordinary frictions. But fleets learn them. They notice where agencies overlap without clarity, where one department assumes another is following up, and where a demand for more information is likely to outlast the season that made the issue valuable in the first place. The stamp matters less than the sequence.

Norway offers one of the strongest examples of capable coastal management in northern waters, and that is precisely why it is analytically useful. Its fisheries administration, coast guard, and associated scientific and monitoring systems are often held up as comparatively effective. That reputation is not accidental. Norway combines licensing, inspection, vessel tracking, scientific input, and a long tradition of fisheries administration in a way many states would envy. Yet even here the limits are obvious once one looks beyond formal authority. Svalbard remains politically sensitive. Weather and darkness constrain what can be done safely. Enforcement at sea must still survive diplomatic scrutiny, legal argument, and evidentiary challenge. A relatively strong system does not abolish the implementation problem. It clarifies it. If a capable state still faces operational and bureaucratic strain, weaker or more fragmented systems will face it more severely.

A second named example sits farther west. The 2018 Agreement to Prevent Unregulated High Seas Fisheries in the Central Arctic Ocean is often treated as a triumph of precaution and institutional foresight, which it was. It is also evidence of coastal states recognising the limits of their own practical reach. The parties did not wait for a full commercial fishery to emerge in a region where monitoring, response, and co-ordination would all have been difficult from the outset. They built restraint and joint process before competitive exploitation could expose the thinness of pure enforcement. The lesson is plain. A wise coastal state does not merely ask what it is entitled to regulate. It asks what it could actually police if incentives sharpened faster than institutions matured.

This is why bureaucratic design deserves as much attention as geography. Analysts often speak of state capacity as though it were a single resource that can be counted in ships, aircraft, or budget lines. In practice, capacity is institutional choreography. Which agency issues the licence? Which verifies the data? Which boards the vessel? Which handles infractions? Which decides whether an anomaly merits sanction? Which ministry owns the diplomatic consequences if enforcement becomes politically sensitive? Fragmentation can arise not because agencies are incompetent, but because their mandates overlap awkwardly or leave gaps between legal authority and operational responsibility. A fisheries agency may know what constitutes a breach. A coast guard may know where the vessel is. A prosecutor may hesitate because the chain of evidence is incomplete. None of this looks like collapse. It looks like governance that is almost enough.

Almost enough is dangerous in fisheries because it creates exploitable ambiguity. Operators do not need to break rules flamboyantly when institutions are fragmented. They need only understand timing, procedure, and burden of proof. A vessel that responds slowly to an instruction, fishes close to a jurisdictional seam, or manages its paperwork so that one discrepancy is always explainable in isolation may frustrate enforcement without inviting a decisive response. Administrative fragmentation lowers the cost of such behaviour. It makes the difference between suspicion and sanction wider. This is especially true where multiple agencies must agree before a case becomes politically or legally meaningful. Grey conduct thrives not only in the open sea. It thrives in the space between offices.

The relevant abstraction here is enforcement efficiency, but it should be made concrete. Efficiency is visible when a suspicious track becomes a timely inquiry, an inquiry becomes a coherent inspection, an inspection produces admissible documentation, and documentation leads to a sanction that others recognise as credible. Inefficiency is equally visible: duplicated requests, conflicting databases, slow legal review, uncertain chains of command, missed boarding windows, and cases that dissolve after detection because no one body owns the full sequence. Coastal states often speak in the language of sovereignty, but fleets respond to these procedural details. They compare states not only by the stringency of their rules but by the reliability with which those rules travel from notice to consequence.

A reasonable counterpoint should be acknowledged. Coastal control is not always undermined by fragmentation, and some degree of functional separation can be beneficial. Distinct agencies can create checks, reduce politicised concentration of power, and bring specialist expertise to different parts of the chain. A scientific body should not become an enforcement agency. A coast guard should not be forced to invent stock assessment. A prosecutor should not act as a fisheries manager. Coordination, not bureaucratic fusion, is the real test. Some states manage these divisions well enough that specialised institutions reinforce one another rather than creating exploitable seams. The problem, then, is not plurality as such. It is poorly connected plurality under conditions where delay and ambiguity are expensive.

That limitation sharpens the core point rather than diluting it. Coastal state power in Arctic fisheries must be judged by implementation design. A state may possess extensive legal rights and still fail because its reporting systems do not match its sanction pipeline, because field officers and analysts work to different timelines, because diplomatic sensitivity discourages follow-through, or because port inspection and at-sea enforcement do not speak to one another. These are not glamorous failures. They are administrative ones. Yet fisheries politics is full of consequences produced by administrative design that outsiders misread as geopolitical drama or simple non-compliance. Often the deeper truth is duller and more unsettling: the institutions were not built to turn authority into action under the pressure they actually faced.

This is why port controls and paperwork matter so much to coastal states, often more than public debate admits. A state that cannot board every suspicious vessel at sea may still shape behaviour through landing obligations, traceability requirements, catch documentation, service denial, and licensing conditions that make later verification possible. These tools do not replace offshore enforcement. They compensate for its unevenness. They also require their own bureaucratic coherence. Port officers must be able to interpret rules consistently. Data must connect across agencies. Sanctions must follow documented irregularities without disappearing into administrative fatigue. The argument returns to implementation because it always does. Coastal states do not manage northern fisheries by force alone. They manage them through files that must still bite when the vessel is long gone.

Domestic politics complicates the picture further. A coastal state may know that tighter co-ordination, stronger reporting rules, or more intrusive verification is institutionally sensible and still hesitate because local fleets resist the burden, ministries compete over responsibility, or wider diplomatic relations make vigorous enforcement unattractive at a given moment. Capacity is therefore political as well as administrative. A law is easier to pass than to sustain against industry pressure. An inspection regime is easier to announce than to defend when it produces the first major dispute. Governments sometimes prefer vague competence over sharp implementation because vagueness leaves more room for selective accommodation. Fleets notice that too.

The quiet motif throughout is the hand-off. One office to another. One patrol to another. One dataset to another. One legal step to another. Coastal state management works when the hand-offs are fast, intelligible, and owned. It weakens when responsibility moves but accountability does not. In the Arctic, where ecological motion and operational delay already favour ambiguity, poor hand-offs do not remain internal administrative flaws. They become part of the fishery's incentive structure. A clever operator needs less brute evasion where institutions already struggle to pass the file cleanly.

The deeper lesson is almost embarrassingly simple. Formal authority is not the same thing as governing power. A state may regulate licences, zones, seasons, and access in law, yet still find that its true ceiling is set by bureaucratic design, evidentiary discipline, and the ability to convert ordinary information into timely consequence. That is why capacity varies not only with wealth or geography but with how institutions are arranged and whether their routines are built for friction rather than for theory. A northern fishery is not managed by the boldness of its statute book. It is managed by the quality of its hand-offs.

The unresolved question, then, is this. When a coastal state claims control over a moving fishery in harsh waters, what should count as real capacity: the legal right to regulate, the patrol asset that can sometimes arrive, or the administrative system that can connect licence, detection, inspection, evidence, and sanction quickly enough that fleets stop treating fragmentation as part of the fishing ground itself?

RFMOs and co-operative governance

In October 2018, in Ilulissat, Greenland, ten parties signed the Agreement to Prevent Unregulated High Seas Fisheries in the Central Arctic Ocean. The immediate result was not a new patrol regime or a dramatic assertion of sovereign force. It was procedural restraint. The parties bound themselves to consultation, joint scientific work, and a shared decision not to rush into commercial fishing where ecological knowledge and enforcement capacity were still thin. That choice captured the essential logic of regional fisheries management organisations and related co-operative bodies. They do not eliminate conflict. They give conflict a room, an agenda, a timetable, and a language other than collision.

That may sound modest. It is not. Fisheries disputes become dangerous not only when rules are absent, but when disagreement has nowhere legitimate to go except into unilateral quotas, public accusation, or selective enforcement at sea. An RFMO offers a different path. It standardises terminology, gathers scientific advice, records objections, formalises reporting, and subjects competing claims to procedure. The file stays open. That matters because a dispute kept procedural is not a solved dispute, but it is often a less combustible one. The table holds.

Regional fisheries management organisations are best understood as bargaining arenas that also make rules. That dual character is their strength and their burden. They are not neutral scientific bureaux floating above politics, nor are they mere diplomatic theatres detached from ecological management. They sit awkwardly in the middle, where evidence, allocation, compliance, and state interest must be processed together. The show is familiar: states arrive with different fleet structures, different dependencies, different readings of scientific uncertainty, and different ideas of fairness. The name for the institutional answer is co-operative governance. The explanation is that shared stocks and high seas fisheries cannot be managed through unilateral authority alone. The implication is sobering. Governance becomes only as strong as the willingness of parties to keep submitting their disagreements to a common procedure.

The practical strengths of these bodies are considerable. They set standards. They harmonise reporting. They define technical measures, observer requirements, inspection schemes, and compliance expectations that would otherwise vary even more sharply across fleets and jurisdictions. They create transparency by requiring states to declare positions and submit information in comparable formats. They also make the argument legible. A quota dispute placed inside an RFMO is no longer only a bilateral grievance. It becomes a matter that must be justified within a recognised process, against a body of shared scientific work, and under the eyes of other members who may not welcome a precedent of procedural collapse. This is not moral magic. It is institutional pressure, and it can be remarkably effective.

The North-East Atlantic Fisheries Commission offers a useful example because it has long operated in waters where valuable stocks, overlapping jurisdictions, and distant-water fleets generate recurring tension. Its influence has not come from omnipotence. It cannot abolish conflict or guarantee perfect compliance. What it does provide is a procedural framework in which standards, control measures, and technical disputes can be handled without every disagreement immediately becoming a test of bilateral resolve. That is a real achievement. When a regional body can keep states talking in the language of objections, recommendations, inspections, and evidence rather than in the language of insult and retaliation, it is already performing security work, whether or not anyone calls it that.

A second example returns to the Central Arctic Ocean agreement, which, though not an RFMO in the mature sense, illustrates the same governing instinct. Faced with ecological uncertainty and the prospect of future exploitation in remote waters, the parties did not wait for a fully formed fishery and then try to govern it through improvised reaction. They built a co-operative frame around absence. That was more than precaution in the narrow conservation sense. It was a recognition that procedural order itself is a strategic asset. Joint science, consultation, and agreed restraint can buy time where direct enforcement remains difficult and where premature competition would harden positions before institutions had a chance to form. The table appears again here, not because it settles the fishery, but because it prevents the fishery from being settled first by the fastest actors.

Yet these institutions carry deep weaknesses, and the most important of them lies in consensus politics. Co-operation sounds admirable until one remembers that every significant change may require the acquiescence of parties whose incentives point in opposite directions. A state benefiting from an older allocation can use procedural caution to delay adaptation. A newly advantaged participant can use the threat of unilateral action to pressure the institution from outside. Scientific advice may support one reading of the stock while political tolerance for redistributing access lags far behind. Consensus, in this setting, preserves legitimacy by giving all parties voice. It can also preserve paralysis by giving all parties leverage.

This is why RFMOs so often lag behind ecological change. Fish move faster than institutional comfort. A stock extends into new waters, the seasonal centre of gravity shifts, or a high seas pocket becomes more commercially meaningful, and the regional body finds itself working with older allocation assumptions, older control measures, and a procedural culture built for slower adjustment. The lag is not always negligence. Sometimes it is the cost of legitimacy. A system that moves too quickly risks losing members to defection or objection. A system that moves too slowly risks losing relevance because the fishery it governs no longer resembles the one its rules describe. Co-operative governance lives inside that tension. It cannot simply choose efficiency without endangering consent.

Uneven compliance deepens the problem. An RFMO can adopt excellent measures on paper and still discover that implementation depends on flag states with different capacities, different enforcement appetites, and different domestic constraints. One member may inspect rigorously and report conscientiously. Another may lag in data quality, legal follow-through, or sanctioning. The body can note this, review it, and sometimes shame it. It cannot always cure it. This is where standard-setting meets sovereign friction. A regional regime can harmonise expectations far more easily than it can harmonise effort. Fleets notice the difference quickly. They learn which flags carry stronger compliance cultures, which ports scrutinise more closely, and which procedural breaches are likely to remain trapped in committee language rather than becoming meaningful consequences on the water.

The result is that RFMOs often govern by reputational and procedural pressure more than by direct compulsion. This is not weakness alone. It is part of their design. They can lower temperature by keeping dialogue alive when bilateral relations are poor. They can channel disputes into objections, working groups, technical reviews, and incremental revisions. They can make unilateral behaviour more visible and, therefore, more costly in diplomatic terms. A state may still defect. But defection from a recognised regional process is different from action in an institutional vacuum. It must be justified, defended, and managed against a common record. Procedure does not prevent opportunism. It raises its price.

A sensible limitation must be admitted. Not every fisheries dispute benefits from being kept procedural if procedure becomes a substitute for adaptation rather than a means of managing it. A regional body can become too comfortable with its own language of meetings, reports, and recurring agenda items. The form of co-operation survives while the substance decays. Members continue to attend, objections continue to be recorded, and scientific papers continue to circulate, yet the institution no longer changes behaviour where it most needs to. This is a real danger. The table can hold a dispute for so long that it begins to preserve it. Institutional continuity, though valuable, is not always evidence of institutional success.

That limitation is especially important in Arctic and sub-Arctic settings because ecological novelty and geopolitical sensitivity can push in opposite directions. A stock shift may call for faster revision of access, but wider political relations may make members less willing to concede anything that looks like precedent. A high seas seam may require tighter collective control, but some parties may fear that stronger measures today will constrain them tomorrow in a different fishery. A regional body must then decide, often implicitly, whether its priority is to adapt rules quickly or to keep members inside the room. This is not a theoretical dilemma. It shapes whether the institution remains ecologically credible, politically durable, or both in imperfect measure.

The quieter achievements of RFMOs are therefore easy to undervalue because they often take the form of non-events. A dispute does not become a detention. A contested stock does not produce an immediate quota war. A technical disagreement remains in a scientific body rather than spilling into public insult. These outcomes are hard to celebrate because they are hard to photograph. But they matter. They show that the procedure can itself be a form of restraint. A fisheries regime need not produce affection to produce order. It may only need to make continued co-operation less costly than a visible breakdown.

The broader lesson is that co-operative governance depends on more than rule texts. It depends on habits of attendance, tolerance for procedural frustration, and a continuing belief that even a disappointing forum is better than none. It also depends on the institutional imagination to revise measures before the fishery outgrows them. RFMOs fail when they become either too weak to matter or too rigid to adapt. They succeed when they hold together enough common discipline that states continue to bargain through them even while distrusting one another's motives. That is not elegant. It is governance.

The quiet motif throughout is the table. States arrive at it with grievances, data, scientific caveats, domestic pressure, and sometimes very little warmth. The table does not erase any of these things. It arranges them. It gives objection a form, delay a record, concession a procedure, and defection a visibility it would otherwise lack. That arrangement is often the difference between a difficult fishery and a dangerous one.

The question that remains is not whether regional co-operation is preferable in the abstract. It usually is. The harder question is what a regional body should protect first when ecological change outruns institutional comfort: the consensus that keeps members at the table, or the adaptive speed needed to keep the table from governing a fishery that has already moved somewhere else.

High seas pockets and regulatory seams

In October 2018, in Ilulissat, Greenland, ten parties signed the Agreement to Prevent Unregulated High Seas Fisheries in the Central Arctic Ocean. The immediate outcome was preventive rather than dramatic: they agreed not to rush into commercial fishing in a region where the ecological picture was incomplete and the enforcement picture even thinner. That decision made sense only if one understood a simple institutional truth. Outside national zones, restraint is harder to verify, harder to enforce, and easier to test. The seam is where patience frays.

A high seas pocket exerts a peculiar political pull because it lies just beyond the place where a coastal state can act with the greatest legal confidence. It may be surrounded by exclusive economic zones, closely tied to stocks that spend most of their lives inside national waters, and yet formally sit outside the most direct line of state control. This produces an asymmetry that fishers, companies, and governments all recognise. The resource may be biologically connected to nearby national systems, but the governing authority is thinner, slower, and more collective. That difference changes behaviour. It makes risk-taking more tempting not because law disappears, but because law has to travel farther and through more hands before it bites.

The central problem is easy to state. Free-riding is most attractive where the gains from early access are concentrated and the costs of sanction remain uncertain. Show a commercially valuable stock moving through or aggregating near a high seas pocket; name the mechanism as a regulatory seam; explain that no single coastal state can simply impose order there in the same way it might within its own zone; and the implication becomes obvious. A vessel can treat the seam itself as opportunity. It can fish where the stock is still biologically tied to neighbouring states' management concerns while counting on the fact that verification, boarding, prosecution, and diplomatic consensus will all be harder once the activity occurs just outside the clearest line of jurisdiction.

The Barents Sea "Loophole" in the 1990s remains one of the most instructive examples. This high seas area in the central Barents sat amid waters of intense importance for cod and other northern stocks. When fish became available there, the legal geography mattered as much as the ecology. The attraction was not only abundance. It was jurisdictional geometry. Fishing in a high seas enclave offered access to stocks closely associated with adjacent coastal management while making direct control more politically and legally awkward. The show was clear enough: fleets moved into a space where incentives to catch were high and the nearest states could not act with the same ease they enjoyed inside their own zones. The name for the problem was not lawlessness. It was seam fishing. The implication was severe. Even a well-managed regional stock could be destabilised if the most tempting part of its range sat where accountability thinned.

A second example, more preventive than reactive, returns to the Central Arctic Ocean agreement signed in 2018. There the parties acted before a full commercial fishery had emerged, precisely because they understood what high seas pockets do to incentives. If a stock of commercial interest were to become available in that central area, parties would face the classic seam problem in a harsher form: remote waters, thin infrastructure, uncertain biology, and no coastal state able to convert immediate concern into straightforward control. The agreement's restraint and research provisions were therefore not merely environmental caution. They were an attempt to deny the seam its usual political logic by building collective discipline before competitive use could normalise itself.

What makes these areas especially difficult is that agreements often exist and still do not settle the practical problem. Regional bodies can set rules, define reporting obligations, establish control measures, and create inspection schemes. These are serious achievements. But verification remains harder when vessels operate far from ports, when patrol opportunities are sparse, when data arrive through multiple states and agencies, and when sanctions depend on the flag state or a wider political process rather than on a single coastal administration with clear command. A fish caught in a seam can therefore take longer to move from suspicion to proof and from proof to consequence. That delay matters. It alters incentives before any formal non-compliance finding is issued.

This is why regulatory seams attract not only outright rule-breaking but more calibrated forms of opportunism. A vessel need not fish in flamboyant defiance of an agreement to benefit from the seam. It may hug the edge of a pocket where oversight is least predictable. It may use legal ambiguity around stock identity, area of capture, or timing of gear deployment. It may rely on the fact that one state's surveillance data must be reconciled with another state's reporting system and a regional body's procedures before anything serious occurs. The abstraction here is risk. Its visible form is delay: extra hours before a boarding decision, extra days before records are compared, extra weeks before a committee turns suspicion into an actionable case. In fisheries, delay is often a kind of permission.

The seam also exposes a deeper institutional mismatch. Fish do not arrange themselves according to legal categories. A stock may spawn largely within national waters, feed partly in a high seas pocket, and be caught by fleets whose obligations differ by flag, by membership of regional regimes, and by domestic enforcement culture. The ecological unit and the legal unit fail to coincide cleanly. When that happens, a high seas pocket becomes more than a geographical curiosity. It becomes a place where responsibility is dispersed just enough for each actor to insist that someone else should carry more of the burden. Coastal states point to the need for collective discipline. Distant-water states point to agreed procedures. Regional bodies point to members' implementation duties. Meanwhile the stock moves through the seam without caring which legal theory best describes it.

A useful limitation should be admitted. Not every high seas pocket turns into a chronic governance failure, and not every vessel operating there behaves opportunistically. Some regional regimes do manage to establish strong enough expectations that free-riding becomes diplomatically and commercially costly even where direct enforcement remains difficult. Procedural discipline, shared science, agreed control measures, and reputational pressure can meaningfully reduce the appeal of testing the seam. This matters because it prevents a lazy determinism. The existence of a pocket does not guarantee disorder. Institutions can blunt the incentive structure. They just rarely erase it altogether.

That limitation, however, reinforces the larger point. The real question is not whether agreements exist, but whether they change behaviour quickly enough under pressure. A control scheme that works tolerably well in calmer settings may prove too slow when a stock suddenly becomes commercially attractive in a remote high seas area. A reporting system may produce transparency after the fact without deterring the initial rush to fish. A regional objection procedure may preserve legitimacy while allowing enough operational delay to reward exactly the conduct it later reviews. Governance in seams is therefore tested at the point where procedure meets tempo. If the seam offers immediate gain and only delayed consequences, it will continue to attract strategic effort.

This is one reason high seas pockets provide such an important bridge to the problem of grey fleets and monitoring. The seam rewards ambiguity. It makes identity, timing, and proof central. Who was fishing there, under what flag, for which stock, with what reporting trail, and under which control scheme? These questions become harder, not easier, outside the zones where a coastal state can integrate surveillance, boarding, port follow-up, and sanctions under one administrative roof. Monitoring technology can reduce uncertainty, but it does not abolish the seam. A track is not a prosecution. A suspicious pattern is not yet an attributed violation. What the seam offers, above all, is room between the visible and the provable.

The political consequences run further than enforcement. High seas pockets put pressure on co-operation itself because they expose asymmetries of cost and benefit. A coastal state may feel that a stock it has historically managed and constrained within its zone is being harvested just beyond its strongest reach. Another state may argue that lawful access under a regional framework is being unfairly stigmatised. A regional organisation may worry that stronger measures will alienate members whose consent it still needs. Each actor can tell a plausible story. The seam sharpens all of them because the underlying grievance is not only about fish. It is about who bears the burden of acting responsibly when jurisdiction is shared and temptation is concentrated.

The quiet motif throughout is the seam itself. Not a dramatic frontier line, but a narrow institutional join where different authorities, expectations, and evidentiary burdens meet. Seams are rarely where governance looks strongest on paper. They are where fleets go to discover how strong it is in practice. That is why these spaces deserve more attention than they usually receive. They reveal whether management systems can carry ecological reality across legal discontinuity without turning every contested catch into a test of patience and precedent.

The unresolved question is therefore sharper than it first appears. When a commercially valuable stock enters a high seas pocket linked to nearby national fisheries, what should count as adequate governance: an agreement that records restraint in principle, a monitoring system that can describe suspicious behaviour after the fact, or a collective willingness to treat the seam not as a loophole to be explored but as a responsibility that no single state can discharge alone?

Joint commissions, working groups, and routine diplomacy

In 1976, Norway and the Soviet Union established the Joint Norwegian-Soviet Fisheries Commission, later the Joint Norwegian-Russian Fisheries Commission, to manage shared stocks in the Barents Sea. The formal achievement was institutional, not theatrical. No dramatic settlement ended all disagreement. Instead, a recurring mechanism was built: meetings, scientific exchanges, technical reviews, and the steady handling of details that would otherwise have accumulated until they became political incidents. That is how much of Arctic governance actually survives. It survives in rooms.

This is easy to underrate because routine diplomacy is visually unimpressive. A patrol vessel, a detention, a ministerial warning, a legal dispute over access - these produce images and headlines. Technical committees do not. Yet fisheries order in the Arctic depends less on dramatic assertion than on repetitive, procedural labour. Officials compare survey methods, adjust control routines, clarify reporting expectations, review anomalies, co-ordinate inspections, and preserve channels through which disagreement can be registered without immediately becoming accusation. These practices are not ornamental to governance. They are its nervous system. When they function, friction is absorbed before it acquires symbolic force. When they fail, even ordinary problems begin to feel like challenges to authority.

The relevant point is not that process replaces power. It is that process that often determines whether power must be used bluntly. Show a disputed stock, a questionable landing record, an enforcement concern, or a disagreement over timing; name the mechanism that handles it as routine diplomacy; explain that working-level contacts and recurring committees allow parties to contest facts, methods, and responses before public escalation sets in; and the implication becomes clear. Process is a security asset. It reduces the chance that a fisheries problem will be forced to declare itself as a sovereignty problem.

The Joint Norwegian-Russian Fisheries Commission remains the strongest example because it has endured through difficult geopolitical weather while managing fisheries that matter materially and symbolically to both parties. Its significance does not lie in the fantasy that it abolished conflict. It lies in something more exacting. It kept the conflict procedural. Shared stocks could be assessed, catches negotiated, scientific disagreements aired, and control matters discussed through a standing mechanism whose continued existence itself lowered the cost of not overreacting. A stock shift or a compliance concern did not have to start from zero. A file already existed. So did the people, habits, and channels needed to keep the file from becoming a crisis. The room remembered.

That phrase matters because routine governance depends heavily on memory. Working groups and joint commissions do more than exchange information. They build institutional recall. They preserve familiarity with the stock, the fleet, the legal arguments, the technical disputes, and the recurrent points of friction likely to matter when conditions worsen. This memory can be invaluable in Arctic fisheries, where seasons are compressed, data are imperfect, and small misunderstandings can metastasise under public pressure. An official who has dealt with a counterpart for years may still disagree sharply with that counterpart. Yet the disagreement occurs against a background of known procedure rather than fresh suspicion. That difference is often the margin between containment and escalation.

Scientific exchanges are one of the quietest and most important parts of this architecture. When states share survey findings, debate model assumptions, review stock behaviour, and compare interpretations in structured settings, they do not eliminate uncertainty. They domesticate it. Uncertainty remains politically consequential, especially where quota or access are contested. But it becomes less explosive if there is an agreed place to argue about it and a common expectation that argument will continue. The show here is mundane: vessel time allocated for joint work, data tables circulated, methods challenged, revisions proposed. The name is technical co-operation. The explanation is that recurring scientific contact turns uncertainty into a process rather than a void. The implication is significant. Parties are less likely to treat every inconvenient result as evidence of manipulation when they helped build the procedure that produced it.

A second useful example lies in the broader co-operative architecture around the 2018 Agreement to Prevent Unregulated High Seas Fisheries in the Central Arctic Ocean. That agreement was precautionary, but its practical force lay partly in its procedural commitments: consultation, joint science, and the creation of a framework for continued interaction before commercial pressure had fully arrived. This is exactly what routine diplomacy does at its best. It builds channels before a confrontation demands them. Such channels may look thin compared with coast guards, patrol systems, and sanction regimes, but they perform a different task. They keep parties engaged in a shared grammar. Once that grammar exists, even a difficult dispute is less likely to jump immediately into maximalist language.

Routine diplomacy also matters because enforcement co-ordination is seldom a self-executing technical matter. States may have neighbouring jurisdictions, compatible interests, and decent laws, yet still struggle if inspectors, analysts, scientists, and diplomatic officials do not work across institutions with some regularity. A suspicious catch pattern in one zone may relate to activity in another. A boarding incident may require immediate factual clarification through more than one chain of command. Port inspections may matter to at-sea suspicions only if records can be reconciled quickly. Joint working bodies make this possible not by eliminating sovereignty, but by creating predictable interfaces between sovereign systems. These interfaces are where implementation becomes believable. Without them, each state remains formally competent and collectively clumsy.

This is why the process should be treated as a form of infrastructure. Roads, ports, and satellites are easy to recognise as enabling systems. Technical committees and recurring bilateral meetings are less obvious, but they do similar work in political time. They reduce transaction costs. They keep records moving. They clarify who speaks to whom when a problem appears. They create standard operating expectations that are hard to improvise under strain. In Arctic fisheries, where distance already slows response, the absence of such infrastructure can be crippling. A disagreement that might have been handled in a working group is instead deferred, misread, or politicised because no trusted routine exists for handling it early.

A credible limitation, however, must be faced directly. Routine channels can become rituals. Meetings can continue long after their participants have ceased to trust one another's good faith. Technical groups can keep producing papers while real political appetite for compromise evaporates.

Information exchange can turn performative, serving mainly to preserve appearances while each side prepares for a harder public line. This is not a reason to dismiss the process. It is a reminder that the process is only as strong as the willingness of parties to let it carry consequences. A committee that cannot influence behaviour becomes a filing cabinet with diplomatic manners.

Trust erodes in precisely this way when routine channels are politicised. A scientific disagreement once handled as a methodological issue becomes read as a national manoeuvre. A control matter, once resolved quietly, is held back for public signalling. A working group once valued for candour becomes constrained because participants fear that any concession, however technical, will later be used against them domestically or strategically. The show changes subtly at first: slower data sharing, tighter scripts, reluctance to acknowledge uncertainty, more procedural objections, less informal problem-solving. The name for this is the politicisation of routine. The explanation is that wider relations or domestic pressures begin to colonise the very channels built to keep narrow disputes narrow. The implication is profound. Once the room is no longer safe for candour, parties fall back on formalism and spectacle.

That erosion is especially dangerous because it often begins before anyone admits that trust is collapsing. Meetings still occur. Agendas are still circulated. Minutes are still approved. But the real work of quiet adjustment diminishes. Officials stop using the channels to solve problems and start using them to position themselves for later argument. Technical precision remains, but strategic reading intensifies. A data request is interpreted for what it signals. A delayed reply is weighed for what it implies. A proposal for joint work is assessed less as science than as leverage. The routine survives in form while losing some of its shock-absorbing content. At that point, process still exists, yet its security value starts to decline.

And yet one should resist the temptation to romanticise a past of untroubled technocracy. Joint commissions and working groups have always been political in the sense that they manage contested value under conditions of unequal power and divergent interests. Their merit lies not in purity, but in usefulness. They give states a venue to behave politically without always behaving explosively. That is no small thing. A working group need not be innocent to be stabilising. It only needs to remain credible enough that parties still prefer it to confrontation when a difficult issue first arises.

The quiet motif here is the room. Not because physical space is inherently meaningful, but because routine diplomacy depends on repeated encounters in bounded settings where argument can remain technical long enough to stay manageable. The room may be a bilateral commission, an enforcement co-ordination channel, a scientific sub-group, or a standing committee under a wider regional arrangement. Its importance lies in what it prevents. It prevents every disagreement from starting in public. It prevents every anomaly from requiring a fresh diplomatic invention. It prevents the loss of working memory that makes each season feel like a first crisis. In fisheries governance, the room is part of the enforcement system, even when no one aboard a vessel ever sees it.

The deeper lesson is that institutions do not only make rules. They organise the pathways through which strain travels. A fisheries system with strong routine channels can absorb surprise, delay escalation, and preserve room for later compromise even when agreement is thin. A system without them relies more heavily on ad hoc contact, legal brinkmanship, and the hope that no incident will arrive before someone has time to improvise a response. That is a poor design for Arctic waters, where time is short and misunderstandings are expensive. Governance requires more than authority. It requires rehearsal.

The unresolved question, then, is not whether joint commissions and working groups are useful. They plainly are. The harder question is what should be done when the very channels built to absorb friction begin to carry it instead - when the room still exists, the agenda still circulates, and the meetings still occur, but trust has thinned enough that routine diplomacy no longer cools the fishery and may even prepare the language of the next dispute.

Chapter 6

Monitoring Technologies and Blind Spots

What does it mean to watch a fishery spread across ice, darkness, distance, and disputed intent? The intuitive answer is technological confidence: more screens, more signals, more images, more data. But the paradox appears quickly. A vessel can be visible without being identifiable, traceable without being truthful, and recorded without being sanctionable. The more one looks, the more one discovers that surveillance does not end uncertainty. It reorganises it.

That matters because Arctic fisheries are often described as though better monitoring were the natural cure for weak enforcement. Sometimes it is. A vessel track can reveal suspicious patterns. A satellite image can narrow the search area. A camera can expose conduct once hidden behind the rail. Yet none of these tools arrives in a vacuum. Each produces arguments as well as information. Was the signal absent because of equipment failure or design? Does the radar contact correspond to the vessel being accused? Who reviewed the footage, under what rules, and with what chain of custody? The screen glows. The case remains hard.

The central tension lies here: monitoring promises clarity, while fisheries politics keeps turning clarity back into interpretation. Tracking systems, remote sensing, onboard observation, and analytic software all increase what can be seen or inferred. But governance requires more than sight. It requires attribution, comparability, admissible proof, and enough institutional confidence that evidence will travel from detection to action without collapsing into dispute. A vessel shown near a boundary is not yet a violator. A gap in transmission is not yet concealment. A suspicious deck sequence is not yet an enforceable case. Blind spots persist not because technology fails absolutely, but because institutions must still decide what the technology means.

This matters now because the same ecological shifts that move fish also move the politics of proof. As stocks redistribute, fishing effort follows into waters where direct inspection is thinner and where competing fleets, states, and regulators are already primed to read conduct strategically. Under those conditions, monitoring systems become more than administrative aids. They become instruments through which parties defend claims, challenge rivals, and test the credibility of rules. A state that cannot patrol constantly may rely heavily on vessel transmissions and imagery. A fleet that understands this may comply visibly where it is being watched and exploit the ambiguity that remains elsewhere. Surveillance then becomes a contest over signals, not merely a replacement for presence.

The stakes reach beyond enforcement in the narrow sense. Monitoring design affects legitimacy. A camera on board, an observer in the cabin, a mandated transmission, a suspicious gap in data, a mismatch between logbook and landing record - these are not only technical matters. They shape whether fishers believe the rules are being applied fairly, whether regulators trust self-reporting, and whether diplomatic incidents can be managed as evidentiary problems rather than public accusations. If a system looks intrusive but weak, it breeds resentment without confidence. If it looks sophisticated but fragmented, it may impress ministers while frustrating the analysts and inspectors who cannot turn information into consequence. Watching badly can be almost as destabilising as not watching at all.

A further difficulty follows from scale. Arctic fisheries are monitored across vast spaces under hard weather and uneven budgets, which means that surveillance is always selective whether governments admit it or not. Detection becomes a matter of triage. Some tracks are reviewed, some are ignored, some are flagged but not pursued. This selective reality is not a scandal. It is the operating condition. But it does mean that every monitoring system rests on choices about what kind of uncertainty will be tolerated and where. Those choices become politically charged when fleets begin to suspect that blind spots are predictable, or when states realise that data abundance can still leave them strategically unsure about who did what, where, and under which identity.

And yet one should resist the reflex that treats technology as either salvation or illusion. Monitoring tools have transformed fisheries governance in important ways. They make some forms of concealment costlier, some enforcement actions better grounded, and some disputes less vulnerable to pure narrative. A vessel that knows its movements are being compared against licences, catch declarations, and port records behaves in a different informational environment from one operating in a near-documentary void. That is a real shift. The limitation is equally real. Tools work through institutions, and institutions inherit weak links: incompatible databases, understaffed analytic units, evidentiary disputes, and legal regimes that may lag behind the very forms of manipulation the tools were purchased to address.

The path ahead follows that uneasy architecture from signal to meaning. It begins with vessel tracking systems that operate simultaneously as safety devices and compliance infrastructure, then moves outward to satellites and other remote sensing tools that can find activity more readily than they can explain it. From there the argument enters the contested terrain of observers, electronic monitoring, and audit practice, before turning to the back-office bottlenecks that often determine whether data become action at all. It closes by confronting the manipulation of identity itself, where spoofing, false flags, and ownership opacity reveal that the core weakness is often institutional rather than technological.

The deeper question is not whether Arctic fisheries can be watched. They can. The harder question is what sort of governance can still act once watching reveals a world of partial signals, disputed identities, and evidence that arrives faster than institutions can absorb it. When the next suspicious vessel track appears at the edge of a stock shift, what will matter more: the quality of the sensor, the credibility of the data trail, or the capacity of the system behind the screen to decide what the signal is worth before the vessel, and the moment, have already moved on?

Vessel tracking systems and compliance signals

In January 2019, when the Russian trawler *Melkart-5* became the focus of a dispute in the waters around Svalbard, the argument was not built from eyewitness memory alone. It rested, in part, on records: where the vessel had moved, when it had been visible, what had been reported, and how those traces could be interpreted by authorities already primed to read the fishery through law, sovereignty, and risk. The visible track mattered. It did not settle the case. A line on a screen is never just a line. It is a claim waiting to be argued.

That is the right place to begin with vessel tracking in Arctic fisheries. AIS and VMS are often described as technologies of transparency, and they are that, up to a point. They allow states, regulators, analysts, insurers, researchers, and sometimes competitors to observe maritime movement at scales and speeds that earlier systems could not sustain. They make concealment harder, continuity more measurable, and routine co-ordination more plausible. Yet the practical meaning of these systems lies not in perfect visibility, but in patterned behaviour. A track is evidence of motion, not of innocence. A gap is an event, not automatically an offence. The signal blinks. Meaning follows later.

AIS, the Automatic Identification System, began above all as a safety tool. Vessels broadcast their identity, position, course, and speed so that other vessels and shore authorities can reduce collision risk and maintain situational awareness. VMS, the Vessel Monitoring System, emerged more squarely inside the fisheries domain, with regulated vessels required to send positional data to authorities at set intervals through channels not designed for open public visibility in the same way. This distinction matters. AIS carries a long tradition of navigational transparency, but it can also be turned off, misconfigured, or manipulated more easily in ways that matter politically. VMS is usually more tightly bound to regulatory obligations and can therefore support enforcement more directly. The two systems overlap in function while carrying different assumptions about audience, intent, and evidentiary weight. Same sea, different logic.

This is why tracking systems should be read as compliance infrastructure as much as location technology. The show is straightforward: a vessel moves into a sensitive area, slows near a boundary, disappears for a period, reappears on a different heading, or loiters in a pattern suggestive of fishing activity rather than transit. The name for what regulators are seeing is behavioural signalling. The explanation is that movement patterns, especially when paired with licensing data, seasonal closures, catch records, and known fishing grounds, reveal choices rather than mere coordinates. The implication is crucial. Compliance analysis begins not with proving everything a vessel did, but with asking what its movement is consistent with and why that pattern deserves closer scrutiny.

That shift from location to behaviour explains why gaps are meaningful. A signal loss can have innocent causes: poor coverage, equipment failure, weather interference, power issues, human error, or technical mismatch at high latitudes where communications conditions can already be awkward. None of these should be dismissed carelessly. But innocence does not make the gap insignificant. In fisheries governance, significance often lies in context. A gap during ordinary transit may matter little. A gap that begins as a vessel approaches a regulatory seam, a closed area, or a moment of high commercial pressure matters more. A gap that repeats in similar settings begins to look like a pattern. The key insight is modest but powerful. Tracking data rarely prove a violation alone. They can reveal that a vessel is living too comfortably inside coincidence.

The North Atlantic and Arctic fringe offer repeated examples of how this works in practice, even where public disputes often focus on other issues. During periods of stress around shared stocks, authorities and analysts have relied on AIS and VMS records to compare declared behaviour with observed movement, especially near sensitive lines or where compliance with closures and access conditions is under question. The 2019 Svalbard dispute matters here not only because of its legal or diplomatic dimensions, but because it reminds us that enforcement judgements in northern waters are increasingly mediated by digital traces. A vessel becomes legible through repeated signals. It becomes suspicious through the timing, interruption, or inconsistency of those signals. The system does not know intent. It does detect storylines.

A second example lies more diffusely in the high seas and seam areas that have already become central to Arctic governance. The 2018 Central Arctic Ocean fisheries agreement was precautionary partly because the parties understood that any future commercial activity in such waters would be difficult to monitor convincingly by presence alone. Tracking systems would inevitably play a large role in whatever compliance architecture followed. That matters because a high seas pocket rewards exactly the kind of ambiguous movement tracking is best at noticing but worst at conclusively adjudicating. A vessel appears near a stock concentration. It alters speed or route in ways consistent with fishing. Its signals go missing briefly. It later lands elsewhere. None of that is a self-sufficient case. All of it is a behavioural signal. In remote waters, tracking often tells authorities where the question is, not yet what the answer is.

That distinction between detection and truth is the most important intellectual discipline in this area. Too much public discussion treats vessel tracks as if they were direct windows into reality. They are not. AIS especially can be absent, noisy, incomplete, or strategically manipulated. VMS is often more dependable from a regulatory perspective, but it too exists within reporting intervals, transmission rules, and state systems that may vary in quality and timeliness. A screen showing a vessel's movement is therefore a representation produced by rules, hardware, communications networks, and administrative processing. This does not make it weak. It makes it contingent. Authorities who forget this risk overclaiming. Operators who understand it may push just hard enough to exploit the gap between what looks suspicious and what can be sanctioned.

The point becomes sharper when AIS and VMS are compared directly. AIS is often publicly accessible in ways that make it invaluable for open-source analysis, maritime research, and broad situational awareness. It is also more exposed to strategic behaviour because its broadcasting logic is tied to navigation and visibility rather than to closed fisheries compliance alone. VMS is typically mandatory for regulated fishing vessels under specific national or regional rules and reports directly to authorities, which gives it a more formal place in enforcement. Yet VMS is not universally shareable, not equally integrated across jurisdictions, and not always available to the wider set of actors who might otherwise help detect anomalies. This creates a useful but underappreciated asymmetry. AIS makes vessels publicly watchable but not always reliably accountable. VMS can make them more accountable but less publicly interpretable. Effective governance often depends on reading both together.

That reading is rarely mechanical. A vessel's speed profile may suggest gear deployment, but currents, weather, avoidance behaviour, or ordinary seamanship can complicate the interpretation. A cluster of movements may indicate fishing effort, transhipment preparation, or cautious navigation in poor conditions. A sudden AIS silence may reflect concealment, but it may also reflect technical or operational noise that only becomes meaningful when compared with VMS, declared activity, nearby traffic, or known grounds. Analysts therefore work not with certainties but with hypotheses of conduct. This is what makes tracking a behavioural signal. It points. It does not close.

The practical consequence is that tracking systems work best when embedded in broader information cycles. A vessel's movements become far more meaningful when paired with licensing status, area closures, catch documentation, observer notes, camera footage, weather data, port calls, and flag-state reporting. One system alone can rarely bear the full weight of accusation. Combined, they can build a pattern strong enough to justify inspection, denial of service, or deeper inquiry. The show, again, is procedural: an anomalous track triggers review, review triggers cross-checking, cross-checking triggers an action. The name for this is data fusion, but the more important explanation is institutional. Tracking systems matter because they start processes. Their value lies as much in what they prompt as in what they depict.

A useful limitation must be admitted. Vessel tracking has greatly improved fisheries governance, but it can also produce a false confidence that seeing more means knowing enough. Authorities may collect rich positional data while lacking the analytic staff to review it intelligently or the legal framework to act on anomalies without corroboration. Analysts may infer fishing from movement patterns that later prove equivocal. Publicly visible AIS records can encourage overhasty judgments by media or advocacy actors who mistake pattern recognition for proof. This matters because trust in monitoring systems weakens when the gap between technological promise and evidentiary reality becomes too visible. A surveillance tool that overclaims can damage the legitimacy of the broader compliance regime.

And yet the answer is not scepticism for its own sake. Tracking systems remain indispensable precisely because Arctic fisheries cannot be governed through direct inspection alone. Vast distance, darkness, cost, and narrow seasonal windows make constant physical presence impossible. AIS and VMS allow authorities to allocate scarce attention more intelligently. They reduce the search problem. They help prioritise patrols, inspections, and audits. They also create ex post records that can discipline behaviour even when immediate response is impossible. A captain who knows that route, speed, and timing can later be compared against declarations is operating inside a different incentive structure from one who expects movement to dissolve into the sea. This change is real, and it matters.

What tracking does not abolish is ambiguity. Indeed, it sometimes sharpens it by making selective visibility newly useful. Grey tactics flourish in the space between total opacity and definitive proof. Operators do not always need to disappear entirely. They may merely need to produce a track that remains arguable: a gap too short to look conclusive, a route consistent with either fishing or transit, a sequence of movements that raises suspicion without closing legal interpretation. This is why tracking should be treated as an index of compliance behaviour rather than as a transparent mirror of conduct. It reveals how ambiguity is managed, not just where vessels are.

The quiet motif throughout is the blink. A signal appears, vanishes, resumes. Analysts lean forward. Regulators compare notes. A blink can be nothing. It can also be the smallest visible sign of a larger strategy. In fisheries governance, meaning often enters at exactly that scale. Not in the grand dramatic absence of all data, but in the slightly mistimed, slightly misplaced, slightly repeated irregularity that suggests a vessel understands how much the system can see and how much it still cannot conclusively say.

The next step in this argument therefore follows naturally. If AIS and VMS show movement as behaviour rather than truth, and if gaps matter because they sit inside wider patterns of intent and opportunity, then the pressing question is not simply how to collect more tracks. It is how to decide, under costly and imperfect conditions, which blink is benign, which is strategic, and which belongs to a form of grey conduct designed precisely to remain just one step short of the certainty that sanctions demand.

Remote sensing and the search problem

On 3 October 2018, in Ilulissat, Greenland, the parties to the Central Arctic Ocean fisheries agreement committed themselves not only to restraint but to a future in which knowledge would have to precede confidence. That commitment carried a hidden admission. No state, however assertive in law, could simply watch such waters continuously with ships and aircraft alone. The monitoring problem was too large, too expensive, too intermittent. Long before any boarding, the first task would be finding the vessel at all. Search comes first.

That fact changes the whole meaning of surveillance in Arctic fisheries. Public discussion often imagines monitoring as a matter of seeing what is happening and then deciding what to do about it. In practice, vast-area enforcement begins much earlier and on humbler terms. Authorities must decide where to look, which sensors to trust, how to compare imperfect signals, and which possible targets justify scarce follow-up. A satellite image, a radar return, an optical scene, a suspected AIS silence, a vessel track that does not fit the season - these are not conclusions. They are invitations to spend attention. The pixel is not the case.

Remote sensing has therefore become indispensable not because it solves fisheries enforcement, but because it changes the search problem. Synthetic aperture radar can detect vessels in darkness and through cloud in ways optical systems cannot. Optical imagery can reveal visual detail when weather and light cooperate. Satellite-based vessel tracking can suggest where legally declared traffic should be. Other sensors may identify anomalies, compare activity patterns over time, or help analysts distinguish routine transit from concentrated fishing effort. The show is technical but simple in purpose: reduce the area of ignorance. The name for this is surveillance triage. The explanation is that no state can inspect everything, so states must first narrow the field of what deserves inspection. The implication is sobering. Enforcement begins with probabilities, not certainties.

This matters especially in the Arctic and adjacent northern seas because distance multiplies the cost of every mistake. A patrol aircraft sent after the wrong target wastes fuel, time, and an opportunity that may not return before weather closes in. A vessel dispatched too late may arrive after the suspected activity has ended or moved across a seam of jurisdiction. An analyst who treats every anomalous radar return as urgent soon exhausts the system. Surveillance, in such conditions, is not a wall of visibility. It is a sequence of bets. Some are good. Some are unavoidable gambles. Remote sensing matters because it improves the odds, not because it abolishes the wager.

Detection is the first gain. Attribution is the harder one. A radar image may reveal that a vessel-sized object is present in a protected area or near a high-value stock concentration. It may do so with remarkable clarity compared with older methods. Yet presence alone does not answer the question that governance ultimately needs answered: who was it? If identity is uncertain, sanction weakens. A vessel may have altered or ceased broadcasting AIS. It may be operating among several others. It may share broad physical characteristics with lawful traffic. Cloud, sea state, revisit timing, and ordinary signal ambiguity all affect what can be inferred. This is the governing distinction between seeing something and attaching it to a legally and diplomatically usable subject. The sea offers detection more readily than blame.

The problem is not only technical. It is institutional. A coastal state or regional body may obtain a highly suggestive image, but what follows depends on how that image enters a chain of action. Is it compared against licence databases quickly enough? Can it be reconciled with VMS, AIS, or logbook records? Does someone have authority to task a patrol or request flag-state explanation? Can the image be retained and interpreted under evidentiary rules strong enough to survive challenge? Remote sensing does not merely produce information. It tests whether institutions know how to move from suspicion to decision. The search problem is solved only partially by finding the object. It is solved more fully when the system knows what to do next.

The 2018 Central Arctic Ocean agreement is again instructive because its restraint reflected not only ecological caution but monitoring realism. Any future commercial fishery in those waters would emerge in an environment where patrol reach, weather windows, and attribution challenges make pure sea-based enforcement a fragile fantasy. Remote sensing would inevitably be a core part of any practical compliance architecture. Yet the agreement also shows the limit of technological optimism. The parties did not pretend that satellites and shared science would make the central Arctic easy to govern. They recognised that technology could narrow ignorance while leaving the harder political work - agreement, evidence standards, institutional follow-through - very much alive.

A second example lies in the broader enforcement environment of the North Atlantic and Arctic fringe, where authorities increasingly combine AIS, VMS, radar imagery, and other remote indicators to identify vessels operating near regulatory seams or in suspicious patterns. The precise facts vary by case, but the governing lesson does not. Remote sensing is most powerful when it identifies anomalies in relation to declared behaviour. A vessel should be broadcasting here but is absent. A declared transit pattern looks instead like repeated slowing on productive grounds. An optical image shows activity where a closure should have reduced it. These are not convictions. They are reasons to spend scarce attention. The sensor does not replace judgement. It disciplines where judgement starts.

This is why enforcement in remote fisheries should be understood as triage under constraints rather than as omniscient control. Triage is a medical word by origin, but its logic applies cleanly here. Authorities sort possible cases by urgency, probability, and consequence. Some suspected activities will be ignored because the signal is too weak or the asset cost too high. Others will be watched but not immediately pursued. A few will be escalated because the combination of location, timing, sensor confidence, and known regulatory stakes justifies intervention. This sounds unsatisfactory if one imagines governance as total order. It becomes more persuasive once one accepts the actual operating conditions of northern seas. Perfect surveillance is not absent because governments are negligent. It is absent because the ocean is large and the budget finite.

A useful limitation should be admitted. Remote sensing can create false confidence. A vivid image or precise radar return can make analysts, officials, or the public feel that what has been detected is already understood. Often it is not. Vessel-sized objects are not self-interpreting. Optical imagery depends on light and weather. Radar depends on revisit patterns and specialist interpretation. Automated detection tools can misclassify. Even when identification seems likely, timing may blur the picture enough that enforcement action requires corroboration. The danger is not only technical error. It is institutional overclaim. A state that treats suggestive detection as settled proof may damage its own credibility when challenged. Watching more does not license saying more than the evidence can bear.

And yet scepticism can become its own mistake. Remote sensing has transformed fisheries monitoring by making concealment costlier and by collapsing search areas that once overwhelmed patrol systems. In northern waters this matters enormously. A patrol vessel or aircraft no longer has to wander as blindly as it once might have. Analysts can compare patterns over time, identify recurrent hotspots, and use sensors to understand where enforcement gaps are likely to be exploited. This changes behaviour on the water. Fleets know that absence of a patrol does not imply absence of notice. The search problem remains. It is simply less forgiving than before.

Weather and cost ensure that remote sensing remains probabilistic. Weather limits optical systems. Sea state affects interpretation. Satellite coverage varies by orbit, tasking, and budget. The most useful imagery may not be available at the decisive moment, or it may arrive too late for direct follow-up. These are not flaws in the crude sense. They are conditions of use. Surveillance choices are therefore always shaped by trade-offs: broad but coarse coverage or narrower and richer imagery, routine monitoring or targeted tasking, public-source visibility or closed official systems, investment in sensors or investment in the analysts who can turn signals into priorities. Capacity lies in those choices as much as in the hardware itself.

This makes back-office competence inseparable from remote sensing. A state may possess access to radar imagery, optical scenes, and tracking feeds, yet still fail if databases are fragmented, formats incompatible, or analytic staff too thin to compare information in time. A remote sensing system is only as good as its capacity to move from image to queue, from queue to assessment, from assessment to response. The quiet drama here is administrative. A suspicious object seen from space matters only if someone can connect it to a vessel list, a licence condition, a known closure, a past behavioural pattern, or an enforcement threshold that justifies action. Otherwise the image becomes a kind of expensive scenery.

This is where remote sensing also meets the politics of identity. Detection can reveal presence while leaving ownership, flag, or operational control blurred. A vessel may switch identifiers, manipulate broadcasts, or operate through ownership structures that complicate responsibility even if the hull itself is seen. The result is not that remote sensing fails. It is that surveillance collides with attribution. The sensor shows an actor-shaped problem; the institution must still decide whose problem it is. Grey tactics flourish in precisely this interval, because operators understand that being seen is not the same thing as being reliably named.

The quiet motif throughout is the pixel. One pixel, then another, a shape on radar, a lightless contact, a vessel-shaped certainty that is not yet a legal one. Fisheries enforcement in remote seas begins there: with imperfect fragments that invite interpretation and compete for scarce response. The pixel matters because it narrows the search. It does not by itself close the argument.

The deeper lesson is that remote sensing changes the meaning of control. It does not make states all-seeing. It makes them selective under better conditions. That is already a major gain. But it also means that governance depends less on the fantasy of complete visibility than on the discipline to manage uncertainty intelligently. A state that understands this will use satellites, radar, and imagery to ask sharper questions, not to pretend it already has all the answers.

The unresolved question, then, is not whether remote sensing can find suspicious activity in Arctic fisheries. It plainly can. The harder question is what kind of enforcement system can live honestly with the fact that detection is abundant, attribution is harder, and response remains expensive - and still decide, before the next vessel slips through cloud, darkness, or legal ambiguity, which signal deserves the patrol, the query, or the case.

Observers, electronic monitoring, and contestation

In December 2010, the North Pacific Fishery Management Council acted to tighten controls on Chinook salmon bycatch in the Bering Sea pollock fishery, and one practical consequence was immediate: the credibility of monitoring moved to the centre of the argument. The issue was no longer only how much bycatch should be tolerated. It was who would say, with sufficient authority, what had actually happened on deck. A fish can be counted several ways. That is the trouble.

The simplest account of fisheries monitoring says that if governments want better compliance, they should put more eyes on boats. In practice, the question is never so clean. Which eyes? Human observers, who can interpret context and record nuance but may be pressured, limited, or absent at the wrong moment? Cameras, which can preserve a visual record but still depend on placement, review, tamper resistance, and rules about what counts as evidence? Logbooks and audits, which structure accountability but rely on truthful self-reporting and administrative capacity strong enough to compare declarations against something independent? Monitoring is therefore not a device. It is an arrangement of trust, burden, and contest over whose account of the catch will govern.

That contest matters especially in Arctic and sub-Arctic fisheries because direct inspection at sea is rare, weather is punishing, seasons are compressed, and mixed catches can turn routine trips into regulatory problems very quickly. In such settings, governments do not merely monitor to gather information. They monitor to decide whether a fishery remains governable under pressure. Show a vessel operating in difficult conditions, name the issue as evidentiary dependence, explain that enforcement increasingly relies on records created on board rather than by constant official presence, and the implication follows at once: whoever controls the credibility of those records gains influence over the meaning of compliance itself.

Human observers were the first durable answer to that problem in many industrial fisheries. Their strength remains obvious. A trained person on board can watch sorting decisions, note gear use, record species composition, identify contextual details that a camera might miss, and respond to irregular events in real time. A human observer can say not only that fish were discarded, but under what conditions, after what discussion, and with what apparent intent. That interpretive power matters. Fisheries are messy. Decks are crowded, species overlap, weather interferes, and the most important events are often ambiguous in ways only a human witness can parse responsibly. The name for this advantage is contextual judgement. The explanation is that compliance is not always visible as a simple image. The implication is that observer programmes can provide a depth of evidentiary texture that no purely automated system yet matches.

And yet observers also sit inside the very social and physical conditions that make fisheries hard to govern. They are guests, and sometimes unwelcome ones, on vessels whose crews live and work in confined spaces under commercial and weather pressure. Safety concerns can limit what they see and when they can intervene. Coverage is expensive and rarely universal. Programmes differ in training, independence, placement rules, and the authority granted to the data they produce. An observer may become the focal point for disputes over interpretation, privacy, or even personal risk. In some fisheries, the question is not whether observers are useful but whether they can be deployed widely enough, safely enough, and credibly enough for the system to rely on them. A witness is never just a witness. A witness changes the room.

This is why camera-based electronic monitoring has been so attractive to regulators and so contested by industry. A camera does not tire in the same way, can operate on trips where placing an observer is costly or impractical, and can create a reviewable record that survives the voyage. In principle, this helps with the problem of discards, high-grading, bycatch accounting, and the verification of self-reported catches. It also changes the politics of proof. A video file can be audited later, compared against logbooks, and used to test patterns rather than relying on a single observer's notes. For remote and high-latitude fisheries, where the opportunity to verify conduct in person may be rare, that is a powerful advantage.

But cameras do not eliminate argument. They relocate it. Was the camera positioned properly? Was the angle sufficient to identify species and quantity? Did weather, spray, icing, darkness, or deck congestion reduce clarity at the decisive moment? Was the footage reviewed by trained staff using consistent standards? How were privacy concerns handled for crews living in effectively monitored workplaces? Was the system tamper-evident, or merely tamper-prone in a more technical way? The common fantasy that electronic monitoring is objective while human observation is subjective misses the point. All monitoring systems are designed, interpreted, and governed. The real question is not whether a camera is neutral. It is whether the whole chain from lens to sanction is credible enough to bear legal and political weight.

The Bering Sea pollock fishery, especially where Chinook salmon bycatch became a central management concern after 2010, reveals why these design questions matter. Once a protected or tightly constrained stock becomes the species that can choke a target fishery, every monitoring method acquires political charge. Fishers may argue that cameras flatten ecological complexity into punitive scrutiny, or that observers overinterpret deck events under difficult conditions. Regulators and affected communities may reply that without reliable onboard monitoring, bycatch limits are little more than aspirational language. Both are arguing about evidence. Neither is arguing only about technology. They are contesting who gets to define the event that matters: a lawful target haul, an avoidable bycatch incident, or a mixed catch made newly explosive by the management regime.

A second example, very different in legal culture but similar in structure, comes from the European Union's landing obligation, fully extended in 2019 to catches subject to quotas in mixed fisheries such as those of the North Sea. The formal ambition was to reduce discarding by requiring fish to be landed and counted rather than thrown back unseen. This immediately raised a monitoring question. If discards were now prohibited or tightly controlled, who would verify what still happened at sea? Human observers could not cover every trip. Cameras promised greater reach but triggered disputes over burden, privacy, cost, and evidentiary sufficiency.

Logbooks alone were plainly inadequate where incentives to under-report remained strong. The landing obligation thus showed something essential: a regulatory reform aimed at fish could rapidly become a dispute about watching people.

Logbooks sit at the centre of this argument because they remain indispensable even in technologically dense systems. Fisheries administrations cannot function without declared catches, locations, times, gear information, and other self-reported data. Logbooks organise the official story of the trip. They are cheap relative to many other forms of monitoring and can be integrated into licensing, quota accounting, and enforcement systems with great administrative value. But their usefulness depends on auditability. A logbook is not credible because it exists. It becomes credible when it can be compared against observer notes, camera records, vessel tracks, landing weights, port inspections, and known fishing conditions. The name here is documentary compliance. The explanation is that most fisheries are still governed by records operators create themselves. The implication is unavoidable. The fight is not between self-reporting and external monitoring. It is over how much of the self-report can be trusted before independent checks must take over.

This is why monitoring regimes become battlegrounds over burden as much as over truth. Vessel owners and crews bear direct costs when they host observers, install and maintain cameras, store and transmit data, or spend time responding to audits. These burdens are not invented. In remote fisheries, equipment failures are harder to fix, connectivity can be patchy, and the practical demands of documenting operations may feel detached from the realities of weather, safety, and commercial time pressure. Regulators, however, bear a different burden: if they weaken monitoring to ease compliance costs, they may end up governing through declarations they no longer fully believe. Both sides can present real grievances. Monitoring is negotiated because both sides are, in different ways, being asked to live with institutional mistrust.

Privacy intensifies that negotiation. A camera on deck is not the same as a satellite watching a hull from above. It records labour, routine, mistakes, and interpersonal life in a confined and hazardous setting. Crews may feel they are being converted into evidence sources at all times, especially where footage review rules are broad or opaque. Regulators, by contrast, may see claims of privacy as attempts to shield discard practices or to keep the evidentiary threshold too high for meaningful enforcement. Neither response should be dismissed. The issue is not whether privacy matters or whether compliance matters more. The issue is how a monitoring regime can persuade those being watched that its intrusions are bounded, purposeful, and not merely a standing presumption of guilt.

A useful limitation should be admitted. Monitoring technology does not always improve legitimacy simply by becoming more comprehensive. In some fisheries, heavy-handed systems can produce resentment strong enough to damage co-operation, especially if crews believe they are being watched more closely than the underlying regulatory design deserves. A badly designed camera regime or a poorly managed observer programme may generate endless disputes without materially improving compliance. It can also overwhelm agencies with data they lack the staff or procedures to review. Watching more is not synonymous with governing better. A system that records everything and acts on little may corrode trust faster than a system that watches selectively but credibly.

This limitation is especially important because electronic monitoring is often sold as a plug-and-play fix. It is not. Cameras require installation standards, maintenance, review protocols, clear thresholds for action, data storage rules, privacy safeguards, tamper detection, and funding models that do not collapse under routine use. Observer programmes require training, welfare protection, safety rules, and lines of independence strong enough to make their presence credible. Logbook systems require integration, audit capacity, and legal clarity about discrepancies. Monitoring design is therefore institutional design. A state that buys equipment without building procedure buys itself only a different form of ambiguity.

The deeper political issue is legitimacy. Legitimacy is visible when those subject to monitoring believe the system is demanding but not arbitrary, intrusive but bounded, and serious enough that compliance by others is also being tested. It is visible when regulators trust the data enough to act on them and when courts or administrative bodies accept the evidentiary chain without constant collapse. Illegitimacy is equally visible: crews treating monitoring as hostile theatre, regulators quietly discounting their own data because they cannot defend the chain of proof, and public disputes over whether an incident shown on camera or recorded by an observer "really" means what officials say it means. Monitoring fails not only when it misses conduct. It fails when nobody agrees what its records are for.

This is why compliance and legitimacy are bound together more tightly than many policy debates allow. A fishery governed by mistrusted monitoring becomes increasingly expensive to police because every irregularity must be argued from scratch. A fishery governed by monitoring seen as broadly fair may still experience dispute, but the dispute begins from a shared understanding that the records are worth taking seriously. That difference is enormous in Arctic and sub-Arctic waters, where every inspection opportunity is costly and every evidentiary shortcut invites later challenge. Onboard monitoring does not simply collect information. It structures the political terms under which information can travel.

The quiet motif throughout is the witness. Sometimes the witness is human, writing notes in a cabin or on a wet deck. Sometimes the witness is mechanical, a camera fixed above the sorting table. Sometimes it is documentary, a logbook entry later checked against weight and track. None is self-sufficient. Each can be pressured, limited, or reinterpreted. Yet governance depends on them because without some witness the fishery falls back into assertion and denial. The problem is not to find a witness without bias. It is to build a regime in which the witness, human or electronic, is trusted enough to matter and constrained enough not to become intolerable.

The harder question, then, is not whether observers, cameras, or audits are the superior technology in the abstract. Each solves something and distorts something else. The real test is what kind of monitoring settlement a fishery can sustain: one stringent enough to make under-reporting and discard games risky, yet legitimate enough that crews, regulators, and political authorities still recognise the resulting evidence as something more than a permanent argument over who was watching whom, and why.

Data integration and analytic bottlenecks

In October 2018, in Ilulissat, Greenland, the parties to the Central Arctic Ocean fisheries agreement committed themselves not only to restraint and research, but to a future in which information would have to travel across institutions before it could matter. That detail is easy to miss. The great fantasy of modern fisheries monitoring is that once a vessel can be tracked, a haul recorded, a camera installed, or a radar contact detected, governance has somehow solved its problem. It has not. It has only moved the problem indoors. The real test begins at the desk.

That is where expensive systems so often lose their force. A state may possess satellite feeds, vessel monitoring records, AIS tracks, electronic logbooks, observer reports, port declarations, and camera footage, yet still fail to act when it matters. Why? Because collecting data is easier than connecting it.

One database uses one vessel identifier, another uses a different one. Catch records arrive in a format the enforcement unit cannot easily compare with surveillance output. A suspicious track is visible in one system, but the licensing condition that gives it meaning sits elsewhere, accessible only through another office. By the time the pieces are reconciled, the vessel has landed, moved on, or slipped into argument. Information has been gathered. Intelligence has not.

This is the central bottleneck in Arctic fisheries governance. The show is familiar: a monitoring system flags an anomaly, a fisheries analyst requests supporting records, an enforcement officer asks for clarification, a legal adviser queries whether the evidence is sufficient, and the case either advances slowly or stalls in ordinary administrative confusion. The name for the underlying problem is integration failure. The explanation is not technological ignorance in any simple sense. It is the institutional fact that surveillance tools are usually acquired one by one, funded under different mandates, run by different agencies, and governed by different evidentiary assumptions. The implication is severe. A state can spend heavily on seeing more while remaining mediocre at deciding more.

This matters especially in Arctic and sub-Arctic fisheries because time pressure is not an abstract inconvenience there. A suspicious event may occur during a short fishing pulse, near a regulatory seam, under weather conditions that make physical follow-up uncertain. A tracking anomaly means little unless it is connected quickly to licence status, closed-area maps, recent declarations, and whatever independent evidence exists about activity on deck or at landing. If analysts cannot do that fast enough, the monitoring system becomes retrospective rather than operational. It may still help produce later knowledge. It may not alter behaviour in the moment that mattered. A live fishery then becomes governed by archival intelligence.

The problem is often misdiagnosed because governments and commentators alike are drawn to visible assets. Patrol vessels are impressive. Satellite imagery sounds advanced. Onboard cameras and digital reporting systems photograph well in ministerial press releases. Back-office competence does not. Yet the back office is where enforcement becomes real or remains rhetorical. A coast guard can detect. A prosecutor can threaten. A ministry can announce. None of this matters much if analysts cannot connect the vessel track to the quota condition, the logbook entry to the landing discrepancy, the observer report to the relevant legal threshold, or the ownership record to the flag that carries enforcement responsibility.

Capacity, in other words, resides partly in clerical architecture. That truth is deeply unglamorous. It is also decisive.

The European Union's struggle to turn the landing obligation into a genuinely monitorable regime offers a useful example. Once the obligation expanded to quota-managed catches, the policy logic became clear enough: fish that might once have been discarded should be landed and counted. But this immediately created an integration problem. Electronic logbooks, quota accounts, port inspections, camera footage where available, observer records, and gear data all needed to speak to one another in a way that could distinguish a reporting error from a discard offence, or a choke-species problem from deliberate concealment. That proved harder than simply mandating the rule. The issue was not only whether data existed. It was whether different parts of the system could compare and interpret them quickly enough to make enforcement credible rather than symbolic.

A second example lies in northern fisheries co-operation where states possess robust scientific and surveillance traditions but still face institutional friction at the point of action. The Norwegian fisheries administration and coast guard are often treated, for good reason, as relatively capable. Yet even such systems must still solve the integration challenge: vessel monitoring data, licensing databases, port records, scientific assessments, patrol observations, and legal follow-through do not automatically align simply because they are all state-managed. A comparatively strong system shows the problem more clearly, not less. It demonstrates that the bottleneck is not merely a function of scarcity. It is a function of how institutions organise information. If a capable state must work continuously to make its systems interoperable, others operating with weaker administrative traditions face a far steeper climb.

This is why fragmented databases deserve attention as a political problem, not merely a technical annoyance. When one agency holds the vessel monitoring feed, another holds the licensing archive, another the inspection reports, and another the sanction history, every suspected breach depends on co-operation that may be slowed by incompatible formats, legal restrictions on sharing, bureaucratic caution, or simple workload. Fleets do not need to know the software architecture in detail to sense the effect. They learn, through experience and industry talk, whether a flagged event is likely to become a case or whether the administrative hand-offs are too slow or too diffuse to sustain pressure. The data may exist. The deterrent does not.

This is where enforcement should be understood as an information cycle rather than as a sequence of isolated acts. Information is generated by surveillance, tracking, observers, cameras, declarations, and inspections. It must then be cleaned, matched, compared, contextualised, escalated, acted upon, and, crucially, fed back into the system so that future monitoring becomes sharper. The cycle matters because a break anywhere along it weakens not only the immediate case but the credibility of the whole regime. If anomalies are detected but never pursued, analysts stop trusting the value of flagging them. If cases are built but sanctions remain inconsistent, operators adjust accordingly. If legal outcomes do not inform future risk models, the same errors repeat. Data without circulation is only stockpiled visibility.

The phrase "analytic bottleneck" therefore names more than a staffing shortage, though staffing is often part of it. It names the point at which information becomes governance only if someone has the authority, skill, time, and institutional backing to turn fragments into a usable picture. This involves judgement as much as software. Analysts must decide which anomalies deserve scarce investigative effort, which patterns are likely noise, how strongly disparate datasets actually corroborate one another, and whether the resulting picture can survive legal or diplomatic challenge. Enforcement is triage, but triage performed inside institutions that often prefer the comfort of data accumulation to the risk of decisive interpretation.

A sensible limitation should be admitted. Not all integration problems are failures of will or competence. Fisheries data are genuinely difficult to standardise. Different tools serve different purposes. Scientific datasets are not designed exactly like enforcement files; safety signals are not created for quota audits; international sharing may be constrained by law, confidentiality, or diplomatic prudence. Some friction is inevitable and, at times, desirable. A system with no distinctions between databases might move faster, but it might also erode due process, blur institutional roles, or compromise data quality. The challenge is not to abolish separation. It is to prevent separation from becoming an excuse for paralysis.

That limitation matters because it prevents a shallow solutionism. The answer is not simply "one giant platform" into which all information is poured. Such dreams often underestimate how legal mandates, evidentiary standards, security requirements, and international politics shape what can be merged and by whom. Nor is the answer merely to hire more analysts, though many systems plainly need them. The real question is architectural.

Which data need to meet, at what stage, under whose authority, for what type of decision? A good information cycle is not one in which everything is centralised indiscriminately. It is one in which the right comparisons can be made quickly enough, by the right people, without destroying the legal integrity of the process.

This is also where legitimacy enters. A state that integrates aggressively but opaquely may gain speed while inviting accusations of arbitrary enforcement or surveillance overreach. Industry actors may resist if they believe incompatible datasets are being stitched together in ways they cannot understand or challenge. Courts or administrative tribunals may grow sceptical if the chain from raw data to official inference is insufficiently transparent. Conversely, a state that prioritises procedural caution at every stage may preserve formal fairness while losing operational credibility because nothing happens in time. Information cycles therefore have politics embedded in them. They are not only technical arrangements. They decide how much uncertainty will be tolerated, how much explanation will be demanded, and how much speed the system can justify before it starts to look careless.

The quiet motif here is the desk. A vessel may be on a screen. A hold may be full. A rule may have been breached. None of that becomes governance until someone at a desk can pull together what different systems know and decide that the pattern is strong enough to justify action. The desk is where the sea becomes administratively real. It is also where expensive monitoring regimes most often reveal whether they were built as coherent systems or as collections of gadgets purchased under separate pressures.

This is why the back office should be treated as part of maritime capacity rather than as support work in the narrow sense. The analysts who compare logs, tracks, footage, licences, and landings are not ancillary to enforcement. They are part of the enforcement vessel's reach even when they never leave shore. Their competence determines whether patrol assets are sent intelligently, whether suspicious conduct is documented coherently, and whether sanctions are grounded in something more durable than intuition. A state with modest physical assets but strong analytic integration may govern better than a state with impressive hardware and disjointed information flows. The reverse is more common than governments like to admit.

The deeper lesson is almost embarrassingly simple. Fisheries monitoring has become good at generating traces and remains uneven at converting traces into decisions. The bottleneck is not always at the horizon. It is often in the junctions between systems, mandates, and people. Once that is understood, capacity looks different. It is no longer measured only by ships, satellites, or cameras. It is measured by whether the information cycle closes - whether observation leads to comparison, comparison to inference, inference to action, and action back to improved understanding. A regime that cannot close that loop may still know a great deal. It will struggle to govern accordingly.

The unresolved question, then, is not whether Arctic fisheries can be monitored more intensively. They can. The harder question is what kind of institution can turn proliferating data into timely, defensible action without drowning in its own systems - and whether governments are willing to invest political attention in the desk, the analyst, and the hand-off with the same seriousness they devote to the more visible hardware that first made the signal appear.

False flags, spoofing, and identity management

In 2016, the European Union adopted tighter rules aimed at improving the official register of fishing vessels, partly because a vessel is never only a hull in the water. It is also a legal identity, a compliance history, a flag-state responsibility, an owner, an operator, a set of radio and satellite signals, and a paper trail that makes sanction possible if something goes wrong. When those layers line up, governance can move. When they do not, the problem is not merely concealment. It is attribution. A vessel can be seen. It may still remain politically slippery.

That is why deception in fisheries should be understood less as cinematic trickery than as institutional interference. False flags, signal spoofing, identity switching, shell ownership, and manipulated registration details do not matter because they are technologically exotic. Most are not. They matter because they widen the gap between detection and consequence. A radar contact may appear. A suspicious AIS track may raise concern. A landing may not fit the declared trip. Yet if the identity attached to that conduct is blurred, contested, or layered through multiple jurisdictions, the sanction pipeline weakens. The name on the stern is not enough.

The simplest deception is also one of the oldest: false or shifting flag presentation. A vessel may operate under a registration that is outdated, misleading, or strategically changed to complicate which state bears responsibility for oversight and enforcement. The show is concrete enough: a vessel appears under one flag in a public-facing system, claims another in a licensing context, or uses reflagging to move into a jurisdictional or compliance environment thought to be more permissive. The name for this is flag uncertainty. The explanation is straightforward. Fisheries governance still relies heavily on flag-state duties, so changing or obscuring the relevant flag alters who must investigate, who may sanction, and who can be embarrassed diplomatically if the case grows public. The implication is immediate. Even where conduct looks suspicious, authorities may hesitate if they are unsure whose vessel they are really dealing with.

This is not simply a legal nicety. Flags organise the whole enforcement chain. They determine reporting obligations, access to regional fisheries bodies, inspection rights under specific regimes, insurance expectations, and the diplomatic path through which non-compliance claims must travel. A vessel that can muddy that identity gains more than camouflage. It gains time. In Arctic and sub-Arctic fisheries, where seasons are compressed and follow-up is already difficult, time is a form of strategic advantage. A case delayed by uncertainty over flag or registry may become politically stale long before it becomes administratively sound.

Spoofing sits nearby but operates more through signal than paperwork. AIS, and sometimes related systems in broader maritime monitoring, can be manipulated so that a vessel appears somewhere other than where it is, appears under a misleading identity, or produces tracks that confuse rather than clarify. The point is not that every odd signal is deception. Equipment errors, signal collision, poor configuration, and communications limitations all occur. But intentional signal manipulation matters because monitoring systems are built to narrow the search problem and to connect observed movement to a legally usable identity. Once the signal itself becomes contestable, analysts are forced to move from location inference to deeper forensic comparison. The screen lies cheaply.

That cheapness is what makes spoofing so institutionally damaging. A coast guard or fisheries administration rarely relies on one signal alone, but signal integrity still matters because it structures what gets checked first and how quickly. A misleading AIS track can waste patrol attention, distort risk analysis, or allow a vessel to build plausible deniability later. The show may be subtle: a vessel identity that overlaps suspiciously with another, a route inconsistent with weather and declared activity, a signal jump that looks physically implausible, or a pattern of disappearance and reappearance that happens near closures, seams, or sensitive grounds. The name is signal manipulation. The explanation is that the system is being fed information shaped not to vanish entirely, but to remain arguable. The implication is that tracking becomes evidence of ambiguity rather than clarity.

A second, quieter form of identity management lies in ownership structure. A vessel may be formally owned by one entity, beneficially controlled by another, financed by a third, and managed operationally through still other companies spread across jurisdictions. None of this is automatically improper. Shipping and fishing have long used layered corporate forms for financing, liability management, and tax reasons. The problem begins when that layering makes accountability too slow or too uncertain to support sanctions, licence actions, or reputational consequences. The vessel is seen. The responsible actor dissolves into paperwork.

This ownership opacity matters especially in fisheries because effective governance depends on more than punishing the captain at sea. It often depends on tracing a pattern of conduct to the operator who profits, the company that directs trips, the beneficial owner who shapes risk appetite, or the financial architecture that makes repeated evasive behaviour commercially tolerable. If all authorities can reach is the most visible but least strategically important actor, the deterrent effect shrinks. A shell company can be abandoned. A corporate network can be rearranged. A flag can be changed. The sanction lands. The structure remains.

A useful example comes from the wider global struggle against illegal, unreported and unregulated fishing, where regional bodies, port states, and non-governmental monitoring efforts have repeatedly encountered vessels whose identity records, ownership trails, and signal histories did not line up cleanly. The details vary across cases and regions, but the pattern is stable: a vessel moves under a legally arguable identity while its commercial and operational control is harder to pin down than the hull itself. This is not unique to the Arctic, but Arctic and sub-Arctic conditions make it more consequential because enforcement windows are narrower and direct inspection is costlier. Identity uncertainty thrives where verification is already under strain.

A second example returns to the monitoring environment around northern seam areas and high seas pockets, where vessels may operate near the edge of multiple legal regimes. In such spaces, even modest identity irregularities can have outsized effects. A signal discrepancy that might be resolved relatively quickly near a major port becomes more serious when the relevant databases sit in different states, different agencies, or different regional systems. A possible false flag or ownership mismatch then does not merely confuse records. It complicates jurisdictional confidence. Who should query the vessel? Which state must be notified? Which registry should be trusted? Which enforcement threshold has been crossed strongly enough to justify escalation? Identity trouble is thus multiplied by institutional distance.

This is why a useful typology of deception matters. Not all identity problems are the same, and policy grows clumsy when it treats them as one dark category. One type is flag instability: the strategic use of reflagging, mismatched registry claims, or vague nationality to blur responsibility. A second is signal deception: spoofed, duplicated, missing, or manipulated transmissions that complicate tracking and comparison. A third is corporate opacity: ownership and control arranged through multiple layers so that sanctions or reputational pressure struggle to reach the real decision-maker. A fourth might be documentary inconsistency, where logbooks, licences, markings, and registry details do not align. These categories overlap, but separating them helps because each stresses a different part of the institutional system. The problem is not mystery. It is misalignment.

That distinction matters for another reason. Authorities often respond to identity manipulation by demanding more technology when part of the weakness lies in administrative design. Better sensors can help detect unusual patterns, but they do little if ownership databases are outdated, registry co-ordination is poor, beneficial ownership rules are weak, or legal procedures for acting on discrepancies are cumbersome. The issue is therefore institutional vulnerability. A spoofed signal is dangerous chiefly where the system lacks fast comparison tools or agreed protocols. A shell ownership structure is dangerous chiefly where licence conditions, financial disclosure, and sanction rules do not reach far enough beyond the bridge. Watching more without tracing better simply produces richer frustration.

A credible limitation should be admitted. Identity anomalies do not always reflect deliberate deception. Fishing vessels operate in hard conditions. Registries are imperfect. Public AIS databases can contain errors. Crews may inherit outdated equipment settings or paperwork discrepancies that matter administratively without proving strategic concealment. Overinterpreting every inconsistency as malicious can damage trust, waste scarce enforcement attention, and create legal vulnerability when accusations outrun evidence. This matters because the line between anomaly and deception is itself part of the governance challenge. The serious task is not to dramatise every irregularity. It is to decide which kinds of irregularity should change how a vessel is treated.

That task points back to attribution. When identity is uncertain, sanctions weaken not only because guilt is harder to prove, but because political responsibility becomes easier to evade. A flag state can question whether the vessel was truly operating under its authority. An owner can blame a charterer. An operator can point to an error in transmitted identity. A captain can claim technical malfunction. None of these defences need be wholly persuasive to be effective. They only need to make decisive action slower and more contestable. In fisheries, where evidence often arrives in fragments and where disputes can escalate diplomatically, contestability is itself a strategic resource.

This is why identity management intensifies disputes even when the underlying conduct might otherwise have remained manageable. A vessel in a closed area may provoke one kind of response if its identity and ownership chain are clear. The same vessel may provoke a much more fraught response if everything about its legal personality is arguable. Authorities then begin fighting two battles at once: what happened, and who it properly belongs to. The second battle can swallow the first. Public narratives follow. States accuse one another of selective enforcement or bad faith. Technical uncertainty turns into diplomatic friction because institutions built to regulate fishing now have to litigate identity under pressure.

The quiet motif here is the label. A vessel carries labels everywhere: name, call sign, flag, registry number, AIS identity, licence, owner, operator. Governance works when those labels point to the same object and the same responsible chain. It weakens when the labels drift apart. The drift need not be total. It only needs to be enough that an otherwise actionable case becomes one more file for later clarification. In remote fisheries, later clarification is often another way of saying never in time.

This is why the sensible response to deception is not technological fascination but institutional repair. Registry systems need to speak to monitoring systems. Licensing must reach beneficial ownership where it matters. Port controls and market due diligence can sometimes do what at-sea surveillance cannot by forcing identities to converge at the moment fish become commercially valuable. Regional co-operation matters because no single state sees the whole ownership or movement picture in seam-heavy fisheries. None of this is glamorous. That is its virtue. The goal is not to outwit every trick. It is to reduce the value of tricks by making identity harder to separate from consequence.

The unresolved question is therefore not whether vessels can still disguise, blur, or multiply their identities. They can. The harder question is what kind of fisheries regime can make those tactics less profitable: one that watches more closely from above, one that traces ownership more deeply onshore, or one that treats every suspicious label, signal, and registry mismatch as part of a single attribution problem rather than as a series of minor irregularities too easy to explain away one by one.

Chapter 7

Grey Fleet Tactics and Incentives

A vessel does not have to vanish to become difficult to govern. It may hold just close enough to a boundary to keep intent arguable. It may answer slowly rather than refuse. It may file the paperwork, but in forms that reveal little until cross-checked too late. It may change its flag, ownership trail, or landing pattern without ever presenting itself as plainly outlaw. The point is not invisibility. It is friction. A file can be made to slip.

That is the world of grey fleet behaviour. Not piracy, not straightforward legal compliance, and not always conduct that can be named illegal in one clean motion. Grey tactics live in the interval between what authorities suspect and what they can prove, between what markets prefer to ignore and what regulators are prepared to escalate. They are valuable precisely because they are calibrated. The operator who pushes too far invites a case. The operator who stays just inside ambiguity may earn a season's advantage, a weaker evidentiary trail, or a precedent that later becomes harder to reverse. Ambiguity pays.

The central tension lies here: fisheries governance depends on categories, while grey behaviour depends on making categories costly to apply. A vessel is either inside a zone or outside it, lawful or not, properly identified or not, compliant or not. Yet on the water, and across the paperwork that follows it, many of the most consequential actions are designed to remain inconveniently intermediate. A captain fishes aggressively near a line without obviously crossing it. A company reflags in ways that look legally available but institutionally opportunistic. A transhipment pattern obscures origin without quite collapsing into an open confession of laundering. Each act says the same thing in a different dialect: make the state work harder.

That matters because grey conduct is not random misbehaviour at the margins of the system. It is often a rational response to the way enforcement, monitoring, and market incentives are actually structured. Where sanctions depend on attribution, operators test identity. Where boarding is costly and politically delicate, they test timing and seamanship. Where supply chains accept opaque origin if documentation remains formally plausible, they test paperwork. Where domestic politics makes regulators reluctant to escalate, they test the emotional temperature of an encounter at sea. Grey tactics therefore reveal the fishery's real incentive map better than official compliance statistics often do. They show where institutions are least willing, least able, or least co-ordinated.

The stakes reach beyond individual infractions. Grey behaviour alters the meaning of order itself. If a rule exists but can reliably be stretched through slow-walking, procedural argument, boundary play, or identity blur, then the rule remains formally intact while practical control weakens. This is especially serious in northern fisheries because distance and weather already favour selective enforcement. A fleet that learns how to turn administrative hesitation into operating room does not need dramatic illegality to reshape the fishery. It only needs enough tolerated ambiguity that risk becomes profitable. The result is corrosive. Law begins to look performative to those who follow it closely and optional to those who can afford good lawyers, flexible logistics, or a more adventurous appetite for uncertainty.

The economic structure of the fishery matters here as much as the legal one. Grey tactics flourish where product can move through opaque supply chains, where at-sea logistics can interrupt the trail from catch to landing, and where the value of a disputed haul exceeds the expected cost of scrutiny. Labour arrangements matter too. Crews may operate under layered contracting systems that complicate accountability and reduce the social cost of pushing harder at the edge of the rules. Transhipment can be a legitimate efficiency or a concealment device depending on how origin, timing, and records are handled. A regulator looking only at the vessel misses half the problem. The more revealing question is where the fish must pass to become money, and how many hands can blur that passage before the paper becomes politically useful again.

And yet one should resist melodrama. Not every opaque logistics chain is illicit. Not every reflagging decision is strategic in the darker sense. Not every slow compliance response at sea is brinkmanship. Arctic and sub-Arctic fisheries operate under genuine uncertainty, high costs, rough weather, and complicated legal geographies. A management system that labels every awkward act as malicious will soon lose credibility with the very people whose conduct it most needs to shape. That limitation is important. It keeps the analysis disciplined. Grey behaviour matters not because everything ambiguous is sinister, but because some ambiguity is cultivated precisely for institutional effect.

The path ahead follows that cultivated ambiguity where it becomes most useful. It begins with the visible forms grey conduct takes and why risk-tolerant operators value manoeuvre more than spectacle. It then moves into the labour systems, transhipment patterns, and at-sea logistics that can turn legitimate efficiency into concealment or at least into doubt. From there the argument tightens around strategic brinkmanship itself, where minor incursions, procedural delay, and legalistic insistence are used not merely to evade a single rule but to test what authorities will really defend when the moment turns uncomfortable.

The harder question is never simply whether a fleet is grey. That label comforts too quickly. The sharper question is what kind of fishery teaches operators that ambiguity is worth more than compliance, and what kind of state response can narrow that profit without mistaking every contested signal, awkward manoeuvre, or inconvenient paper trail for a crisis worthy of immediate escalation.

What "grey" behaviour looks like

In January 2019, the Russian trawler *Melkart-5* became the centre of a dispute in the waters around Svalbard after a fisheries inspection and the vessel's subsequent handling raised sharp questions about compliance, authority, and intent. The legal arguments mattered. So did the seamanship. But the deeper lesson lay in something less dramatic than open defiance. The episode illustrated how conduct can be made difficult to classify even when it is easy to suspect. A vessel need not flee, falsify everything, or brazenly reject orders to create institutional strain. It may only need to keep its behaviour arguable. The line stays visible.

That is the starting point for understanding grey behaviour in fisheries. Grey conduct is not simply illegality with better public relations. Nor is it ordinary compliance seen through a paranoid lens. It occupies the interval between clear violation and uncontested obedience. A vessel fishes aggressively near a boundary without obviously crossing it. A company files paperwork that is formally complete yet materially unhelpful. An operator changes flag, ownership chain, or landing pattern in ways that remain technically defensible while making accountability slower and costlier. Grey tactics thrive because institutions prefer categories. Grey conduct makes categories expensive.

This matters in Arctic and sub-Arctic fisheries because enforcement is already burdened by distance, weather, fragmented authority, and evidentiary delay. Show a patrol asset that cannot be everywhere, a monitoring system that detects more than it can conclusively prove, and a sanction process that depends on attribution surviving legal and diplomatic scrutiny; name the resulting operating style as strategic evasiveness; explain that vessels and firms can profit by exploiting the gap between suspicion and action; and the implication becomes clear. Ambiguity is not a side effect. It is a resource. For risk-tolerant operators, the ideal outcome is often not invisibility but contestability.

Aggressive boundary fishing is one of the clearest examples because it often appears ordinary until one pays attention to pattern. A vessel works repeatedly along the edge of a closed area, an exclusive economic zone, or a regulatory seam where fish are dense and oversight is politically sensitive. The conduct may remain just inside a plausible interpretation of legality. Gear is set or hauled close enough to raise concern, but not always close enough to support a straightforward case. The name for this tactic is edge exploitation. The explanation is simple. Boundaries organise law more crisply than they organise fish. Operators therefore gain by placing effort where ecological value spills across legal confidence. The implication is that every subsequent decision - whether to inspect, warn, board, escort, or merely record - acquires strategic weight beyond the catch itself.

This behaviour becomes even more useful when authorities fear that decisive intervention will be costly or ambiguous. A boarding in rough weather near a sensitive line may be lawful yet politically awkward. A warning issued too early may look arbitrary; too late, it looks weak. Operators know this. Grey fishing therefore often consists not in crossing the line outright, but in teaching the state that every attempt to police the line will be disputed in fact, law, or optics. The vessel is not merely catching fish. It is testing appetite for friction. A line on the chart becomes a conversation about institutional nerve.

Paperwork manipulation operates in a quieter register, but its effects can be just as serious. Fisheries governance depends heavily on documents: licences, catch declarations, transhipment notices, landing records, quota balances, crew manifests, ownership statements, and reporting logs. Most systems assume that if paperwork is broadly accurate and timely, enforcement can focus scarce resources on the truly suspicious. Grey operators exploit that assumption by producing records that satisfy formal requirements while frustrating clear understanding. A declaration may be filed in time but broken into categories that conceal meaningful species composition. A landing may be reported in a way that is technically defensible while obscuring where the fish were really taken or how mixed the catch actually was. A quota transfer may be documented correctly while the commercial arrangement behind it makes effective control difficult to see.

The show is administrative, but the politics are sharp. Regulators rarely lose cases because no documents exist. They lose them because documents exist in forms that delay comparison, scatter responsibility, or preserve enough deniability that stronger sanctions become hard to defend. The name for this is procedural opacity. The explanation is that modern fisheries management is built on paperwork as much as on patrols, and paperwork can be shaped to satisfy form while weakening substance. The implication is that compliance ceases to be a simple question of whether a report was filed. It becomes a question of whether the report was designed to inform or merely to survive scrutiny.

A useful example emerges in disputes over mixed-stock fisheries and discard controls, especially under regimes where landings, bycatch, and quota balances must be reconciled across several systems. The European Union's landing obligation, extended fully in 2019 to quota-managed catches in mixed fisheries, did not merely increase the need for honesty. It increased the value of strategic explanation. Once unwanted fish could no longer be discarded as casually or invisibly as before, operators had stronger incentives to present catches in ways that reduced apparent non-compliance or delayed the moment when choke-species problems became undeniable. This did not mean that every discrepancy was deceitful. It did mean that paperwork became an active terrain of contest. A logbook entry was no longer just a record. It was an argument prepared in advance.

Opportunistic reflagging is another classic grey tactic because it shifts the burden of accountability without always requiring a clearly unlawful act. A vessel may change flag to enter a more permissive regulatory environment, to complicate which authorities bear responsibility for oversight, or to exploit differences in reporting culture and sanction practice. In formal legal terms, reflagging can be entirely permissible under certain conditions. That is what makes it useful. The tactic's power lies not in illegality but in asymmetry. One flag state may inspect more rigorously than another, share data more readily, or co-operate more swiftly with regional bodies and coastal authorities. By moving into the administrative comfort zone of a weaker or slower system, an operator may gain room to behave more aggressively elsewhere.

This matters because fisheries governance still depends heavily on flag-state performance. A coastal state can detect, query, and complain. A regional body can record and review. But if the relevant flag state proves reluctant, fragmented, or procedurally weak, the case may soften long before it reaches sanction. Opportunistic reflagging therefore allows operators to arbitrage not only legal rules but institutional stamina. The vessel's physical conduct may remain the same. Its administrative shield improves. That is a grey gain of the highest order: not concealment of the act, but dilution of the response.

A second named example comes from the broader North Atlantic disputes over shifting pelagic stocks such as mackerel after 2010. Those disputes are usually analysed in terms of quotas, zonal attachment, and scientific disagreement, rightly enough. But they also revealed the value of conduct that stayed just inside the arguable. States and fleets did not need to reject regulation wholesale to improve their bargaining position. It was often enough to fish assertively in newly favoured waters, adopt unilateral measures defended in legalistic terms, and let institutions struggle to determine whether adaptation had become opportunism. The fish were real. So was the ambiguity. Grey behaviour in such settings is often nested inside larger political disputes, which makes it harder to isolate and easier to normalise.

This is why market structure belongs inside the analysis. Grey conduct is not only produced by weak enforcement. It is also produced by value chains willing to tolerate uncertainty where supply remains profitable and documentation remains superficially usable. A buyer, processor, or intermediary does not need to endorse evasive behaviour openly to make it worthwhile. It need only accept fish whose origin, species mix, or legal status remains arguable rather than settled. When market actors demand only enough paperwork to preserve commercial comfort, they enlarge the operating room for grey tactics upstream. The ambiguity at sea survives because ambiguity can still be sold ashore.

A sensible limitation must be admitted. Not every awkward or suspicious act at sea is grey conduct in the strategic sense. Vessels work in dangerous, fast-changing conditions. Signals fail. Documents contain errors. Boundaries and closures can be operationally confusing. Reflagging can have routine commercial explanations. Overreading every irregularity as deliberate evasiveness would damage trust, waste scarce enforcement capacity, and make regulators look more theatrical than disciplined. This matters because grey behaviour is defined not by mere uncertainty, but by patterned use of uncertainty. The important question is not whether one event is awkward. It is whether a series of acts forms a recognisable operating style that consistently extracts value from institutional hesitation.

That caution helps refine rather than weaken the concept. Grey behaviour should be understood as calibrated conduct that makes proof, attribution, or escalation disproportionately costly relative to the gain being sought. Sometimes the gain is fish taken close to a boundary. Sometimes it is extra time before a bycatch limit bites. Sometimes it is a better position in future quota arguments because present activity has been normalised. Sometimes it is simply marketable ambiguity. The tactic changes. The logic does not. Stay inside argument. Force the state to choose between expensive action and tacit tolerance. Profit from the gap.

The quiet motif here is the line. Not only the line on the chart, but the line between legal and arguable, between visible and sanctionable, between ordinary administration and political incident. Grey operators do not always cross that line plainly. They work it. They learn how close they can get before a patrol is sent, a case is built, or a buyer turns cautious. Their success depends less on stealth than on institutional mood. That is why grey conduct tells us as much about governance as it does about those who evade it.

The deeper lesson is that ambiguity has market value where enforcement is selective and compliance depends on paper as much as on presence. Fisheries systems often speak as though the relevant categories were stable: lawful or unlawful, compliant or non-compliant, transparent or hidden. Grey behaviour exposes the porous middle. It shows how much depends on interpretation, timing, and the willingness of institutions to turn suspicion into consequence. A fishery full of grey tactics is not one where law has vanished. It is one where law is being converted, piece by piece, into a test of how much uncertainty it can bear before it stops governing cleanly at all.

The unresolved question, then, is not whether grey behaviour exists. It plainly does. The harder question is what kind of regime can reduce the profitability of living in the arguable middle: one that draws harder lines and risks overreach, one that improves documentation and tracing until ambiguity loses value, or one that accepts some residual greyness as the unavoidable price of governing mobile fish in distant seas without turning every contested move into a crisis.

Labour, transhipment, and at-sea logistics

In June 2014, the *Thunder*, one of the best-known vessels on INTERPOL's fisheries wanted list, was pursued across oceans after years of operating through false identities, evasive logistics, and opaque commercial arrangements. Its notoriety makes it a poor average case and a useful warning. The lesson was not simply that one vessel broke rules at scale. It was that the vessel's power derived from organisation as much as from audacity: fuel, crew, paperwork, rendezvous, and outlets for the catch all had to function for illegality to become profitable. A fish can be hidden in motion. It can be hidden more effectively in logistics.

That is the right frame for labour and transhipment in Arctic and sub-Arctic fisheries. These subjects are often treated as secondary to quotas, patrols, or surveillance. They are not secondary. They determine whether a fishery can be operated continuously, whether catches can be separated from their point of origin, and whether regulators must chase vessels one by one or can intervene at narrower points where fish, labour, fuel, documents, and value all converge. Logistics is where efficiency becomes governable, or ceases to be. The crate travels farther than the net.

Transhipment sits at the centre of this problem because it is both commercially rational and institutionally dangerous. In legitimate form, it allows fishing vessels to offload catches at sea to refrigerated cargo vessels, avoid long and costly returns to port, preserve product quality, and continue fishing during short or distant seasons. In remote northern waters, where port access is sparse and steaming time is expensive, that logic is powerful. The show is straightforward: one vessel keeps catching while another gathers, chills, and carries the product onward. The name is operational efficiency. The explanation is that at-sea transfer lowers costs and extends effective fishing time. The implication is double-edged. The same transfer that improves lawful productivity can also weaken the chain of verification that tells regulators where fish came from, when they were caught, under what conditions, and in what quantity.

That weakening matters because origin is not a decorative fact in fisheries governance. Origin determines which quota was used, which zone's rules applied, which flag state's responsibilities were engaged, which bycatch limits mattered, and whether the catch can later be treated as lawful in port and market systems. Once fish move from one vessel to another at sea, especially in remote or poorly observed waters, the connection between catch event and commercial product becomes more fragile. Documents may preserve it. They may also simulate it. A catch declaration can accompany a transhipment and still leave enough ambiguity about timing, species mix, or area of capture to frustrate later enforcement. The result is not always invisibility. More often it is dilution. The fish arrive with a story that is harder to test than if they had come straight to port.

This is why regulators have long treated transhipment with a mixture of acceptance and suspicion. Many fisheries could scarcely function at industrial scale without it. Some would become substantially less profitable or less safe if every vessel had to steam back to harbour with every hold. Yet the very features that make transhipment useful to lawful operators make it attractive to risk-tolerant ones as well. It extends time on the grounds. It reduces port scrutiny. It creates a moment when product, paperwork, and identity can be rearranged in ways more difficult to inspect than a simple landing. The name for this problem is chain-of-custody instability. The explanation is that each additional transfer point creates another opportunity for concealment, laundering, or plausible deniability. The implication is obvious to anyone trying to police remote fisheries: logistics can protect the fishery or dissolve its traceability.

Labour belongs inside the same analysis because labour arrangements shape how logistics are used and how much pressure exists to stretch rules. A vessel's crew is not merely a technical input. It is part of the enforcement environment. Long voyages, layered subcontracting, precarious contracts, and weak oversight can make crews more vulnerable to coercion, less likely to challenge dubious instructions, and more dependent on operators whose incentives are set elsewhere. In such settings, a legalistic distinction between "the vessel" and "the labour on board" can hide the real organisation of risk. The show may be ordinary: crews changing through carrier vessels, contract structures that distance beneficial owners from day-to-day responsibility, multilingual chains of command, work routines that leave little room to question documentation or sorting decisions. The name is labour opacity. The explanation is that weakly protected labour often accompanies weakly traceable logistics because both favour control from a distance. The implication is that compliance systems focused only on hulls and holds miss the human architecture through which concealment is normalised.

A second named example, less sensational than the *Thunder* but more representative of regulatory response, lies in the long development of port state measures and documentation requirements across multiple fisheries regimes after the adoption of the FAO Port State Measures Agreement in 2009. The agreement did not outlaw transhipment as such. It recognised that many enforcement gains could be made by concentrating scrutiny where fish become commercially useful: at port entry, landing, service requests, and documentation review. That was a profound administrative judgment. Rather than imagining that every vessel could be followed continuously across vast waters, states began building systems to test origin, authorisation, and record consistency at choke points. The insight was practical and unromantic. You do not need to be everywhere if you know where value must eventually queue.

This focus on choke points is one of the most important institutional responses to the logistics problem. A choke point is any place where the movement of fish, fuel, labour, or paperwork narrows enough that scrutiny becomes comparatively cheap and leverage comparatively high. Ports are obvious examples, but not the only ones. Refuelling arrangements, carrier-vessel rendezvous, service requests, insurance checks, buyer documentation, and cold-chain handovers can all function as moments where dispersed uncertainty is forced into a more inspectable form. The show is simple: a vessel or cargo that has operated across enormous space must, sooner or later, pass through fewer doors. The name is strategic concentration. The explanation is that governance works better where mobility meets bottleneck. The implication is why port controls so often matter more than patrol rhetoric admits.

Yet choke-point governance has its own limits. It often arrives late in the chain. By the time fish reach a port or processor, the ecological and legal event that should have been governed has already happened. Regulators may now deny entry, seize product, or start a case, but they cannot undo a discard, reverse an illegal catch, or reconstruct every detail of a contested transhipment merely by reading paperwork after the fact. Choke points are therefore not substitutes for all other monitoring. They are compensations for the impossibility of total at-sea control. This distinction matters because policy discussions sometimes praise port-based leverage as though it resolved upstream opacity completely. It does not. It narrows the space in which opacity remains commercially useful.

A useful limitation must be admitted here. Not all transhipment indicates concealment, and not all complex labour arrangements imply exploitation or intentional evasion. In many distant fisheries, at-sea transfer is a rational response to geography, seasonality, and market timing. Carriers can reduce fuel waste, improve crew safety by shortening some risky return legs, and preserve catch quality that would otherwise suffer. The serious analytical error is to moralise logistics in the abstract. The real issue is whether the fishery's institutional design can separate legitimate efficiency from concealment with enough confidence that lawful operators are not punished simply for working far from port. That is harder than slogan-level debates about "banning transhipment" usually admit.

Still, the distinction between lawful efficiency and illicit concealment cannot be left to rhetoric. It depends on evidentiary design. Were transhipment events authorised and reported in time? Are there reliable records of which catches moved, between which vessels, under whose flag, in which area, and in what quantity? Can carrier-vessel patterns be matched against fishing activity and later landings? Are labour arrangements transparent enough that responsibility travels with the product rather than dissolving into subcontracting? These questions turn logistics into governance rather than leaving it as background infrastructure. Where answers remain weak, the fishery drifts towards grey practice whether or not any one event can be easily named illegal.

This is why supply-chain opacity matters so much. A fish does not cease to be politically important once it leaves the net. It moves through holds, carrier decks, refrigeration systems, customs codes, sale documents, and processing plants, each stage offering opportunities either to preserve traceability or to blur it. Supply-chain opacity is not only a problem for consumer ethics or certification schemes. It is an enforcement difficulty because it separates the point of commercial recognition from the point of capture. The farther apart those points are, institutionally and geographically, the more likely it becomes that no one actor sees enough of the whole chain to intervene decisively. That separation is a logistical advantage for operators comfortable with ambiguity.

One reason regulators increasingly focus on transhipment and labour is that both sit at the intersection of market structure and state capacity. A fleet operating in remote northern waters may be governed less by the threat of being boarded on every trip than by whether its catch can be landed, sold, insured, serviced, or certified without raising questions. Labour abuses, carrier-vessel dependencies, and weak documentation do not merely offend regulatory ideals. They expose where a business model depends on a fragmented accountability chain. Once that dependency is visible, choke-point governance becomes more rational. It targets not only the moment of possible illegality, but the architecture that makes questionable conduct scalable.

The quiet motif throughout is the handover. Fish change hands. Documents change hands. Crew responsibilities change hands. Fuel, risk, and accountability change hands too. Every handover can be legitimate. Every handover also creates a chance for origin and responsibility to become less distinct. That is why at-sea logistics deserve to be read politically rather than merely commercially. They tell us where the fishery becomes easier or harder to verify, and who profits when verification weakens.

This also clarifies why "chasing every vessel" is such a poor fantasy of control. In Arctic and sub-Arctic fisheries, the sea is too large, weather too harsh, and effort too mobile for comprehensive vessel-by-vessel scrutiny to be the main logic of governance. States choose choke points because those are the places where dispersed mobility is forced into administrative contact. A carrier requests entry. A refrigerated cargo vessel arrives with mixed consignments. A processor demands documentation. An insurer asks questions. A port denies service. These moments are less glamorous than an at-sea interdiction. They are often more consequential. They turn logistics against opacity rather than letting opacity live comfortably inside logistics.

The unresolved question is therefore not whether transhipment and complex labour arrangements should exist. They already do, and many fisheries rely on them. The harder question is what kind of regulatory settlement can preserve their lawful efficiency while stripping away the concealment value that comes from remote handovers, layered contracts, and product moving farther from the moment and place where it was actually taken - and whether that settlement can be built strongly enough that operators see choke points not as optional paperwork hurdles, but as the real place where ambiguous fishing becomes vulnerable.

Strategic brinkmanship at sea

In January 2019, the Russian trawler *Melkart-5* became the centre of a fisheries dispute in the waters around Svalbard after a Norwegian inspection escalated into a wider argument about jurisdiction, compliance, and the handling of a vessel under contested conditions. The visible event was dramatic enough: boarding, refusal, escort, protest. Yet the real significance of the episode lay in its calibration. The vessel did not need to stage a spectacular breach to create pressure. It needed only to test how far routine non-co-operation could go before authorities were willing to turn a fisheries matter into a larger political confrontation. A minute matters here.

That is the essence of strategic brinkmanship at sea. It is not reckless lawlessness. It is measured pressure. A vessel, fleet, or operating company pushes just hard enough to explore the reluctance of regulators to escalate. It delays rather than denies. It edges rather than charges. It argues procedure rather than rejecting authority outright. The value of such conduct lies in the response it provokes. If the state hesitates, a precedent is created. If the state acts forcefully, the operator may still gain by portraying ordinary enforcement as overreach. Brinkmanship is therefore less about winning a single encounter than about altering the future operating environment through carefully managed friction.

This makes it a fisheries problem as much as a diplomatic one. The show is often modest: a vessel fishes close to a line, slows its response to radio instructions, disputes the interpretation of a boarding order, insists on a favourable reading of a permit, or continues a contested activity long enough to force the authorities into an uncomfortable choice. The name for this is calibrated non-compliance. The explanation is that states, especially in remote or politically sensitive waters, do not like to escalate quickly. They weigh safety, evidence, diplomatic consequences, domestic optics, and the possibility that a hard response will itself become the story. The implication is obvious to risk-tolerant operators. Reluctance is a resource.

The Arctic and sub-Arctic are unusually fertile ground for this behaviour because enforcement already operates under heavy constraints. Weather limits boarding. Distance slows reinforcement. Monitoring may indicate something suspicious without conclusively proving it. Jurisdiction may be clear on paper and sensitive in practice. A fisheries patrol therefore works inside a zone of hesitation even when the law itself is not vague. Operators who understand that hesitation do not need to disappear entirely or falsify everything. They only need to behave in ways that make the next step by the state feel costly. Strategic brinkmanship is, in this sense, an exploitation of institutional prudence.

Minor incursions are one of its most common forms. A vessel may work extremely close to a closed area, a boundary, or a protected seam where fish density makes even marginal access highly valuable. The incursion may be real but brief, or arguable because of drift, timing, or gear interpretation. The operator then relies on the fact that proving intent or sustained violation will be harder than detecting suspicion. The point is not necessarily to secure a huge illegal catch in a single act. The point is to discover how much room exists between detection and decisive consequence. That information is valuable in itself. Once learned, it spreads quickly through fleets and agents.

Slow-walking instructions is another tactic, and often the more effective one. A vessel told to alter course, prepare for inspection, clarify documents, or cease a particular operation may comply, but only gradually, legalistically, and in ways that consume time. Time matters at sea. A delayed response may allow a haul to be completed, a position to change, weather to deteriorate, or the legal situation to become more contestable. The show can look almost mundane: repeated requests for clarification, narrow readings of wording, prolonged equipment checks, cautious manoeuvres presented as safety necessities. The name for this is temporal obstruction. The explanation is that delay does not have to defeat authority to weaken it. The implication is that a state may emerge with technical compliance and practical loss at the same moment.

This is where legalistic argument becomes a tactic rather than simply a defence. Fisheries law is detailed, layered, and often spread across domestic regulation, bilateral arrangements, regional measures, and the general law of the sea. That complexity gives sincere actors many reasons to proceed carefully. It also gives calculating operators many opportunities to insist that each contested act rests on an unresolved procedural issue. A captain or company does not need to prove the broad correctness of its position in the moment. It may only need to establish enough legal friction that the enforcement authority pauses, consults, or backs away from a stronger move. The law is not being rejected. It is being used as drag.

The *Melkart-5* case remains useful precisely because it shows how an apparently discrete inspection can slide into a larger contest over institutional appetite. The issue was not simply whether one vessel had complied. It was whether fisheries enforcement in those waters would be treated as a routine administrative matter or as something already close to a diplomatic confrontation. That distinction shaped every subsequent choice. A patrol authority weighing whether to press harder was also weighing how the next step would be read beyond the fishery. Brinkmanship works when it can force that double calculation early.

A second example comes from the wider history of northern seam disputes and aggressive boundary fishing, where operators have repeatedly tested how closely they can work to a line before states move from watchfulness to intervention. The details vary by fishery and jurisdiction, but the pattern is stable. A vessel or fleet learns that activity just inside the arguable zone often produces more warning than sanction, more paperwork than interdiction. Once that lesson takes hold, the fishery itself changes. What was once a limit becomes a working edge. The state may still insist that the rule has not changed. On the water, the precedent already has.

This is why micro-behaviour matters so much. Minor incidents are not always minor in effect. A single delayed inspection, a tolerated edge-fishing episode, or a radio exchange that ends without strong follow-through can become a reference point for future conduct. Fleets and companies do not need formal legal victories to draw conclusions. They observe which instructions are really enforced, which conditions can be negotiated at sea, which seam generates only symbolic warning, and which authority is reluctant to create a diplomatic problem over a fisheries one. Precedent in such settings is often behavioural before it becomes legal. The rulebook remains printed. The operational rule has already shifted.

A useful limitation must be admitted. Not every awkward encounter at sea is strategic brinkmanship. Vessels operate in bad weather, complicated currents, mixed jurisdictions, and stressful working conditions. Safety-driven caution, translation difficulty, communication delay, and genuine legal uncertainty can all produce behaviour that looks evasive from shore or from a patrol deck. Overreading every contested manoeuvre as calculated brinkmanship would be both analytically lazy and politically dangerous. States that do so risk escalating incidents that might have been managed calmly and undermining the legitimacy of their own enforcement posture. The concept is useful only when applied with discipline.

That discipline depends on pattern, not isolated suspicion. Strategic brinkmanship is best identified where behaviours recur in forms that consistently test the same institutional weaknesses: repeated edge work near boundaries, habitual delay in response to instructions, selective invocation of legal ambiguity, or conduct that becomes more assertive precisely where authorities face the greatest operational or diplomatic constraints. The key question is not whether one act appears awkward. It is whether a series of acts reveals a deliberate method of converting state caution into operational room. When the same operator, fleet segment, or regulatory environment repeatedly generates such encounters, the burden of interpretation shifts.

This is why market structure and fleet organisation matter here as much as seamanship. A vessel operating under heavy debt pressure, narrow seasonal opportunity, or a commercial culture that rewards risk may be more likely to adopt calibrated non-compliance if the expected gains exceed the expected costs. A company with legal support and experience in contested waters may be especially willing to push procedural edges because it knows how to turn a maritime incident into a bureaucratic argument later. Brinkmanship at sea is therefore not only personal bravado. It is often a business model expressed through manoeuvre.

The state's problem is equally structural. Authorities must decide not only what the vessel has done, but what response will shape future incentives best. A weak reaction may normalise assertive conduct. An overly forceful one may hand the operator a claim of victimhood, harden diplomatic relations, or make ordinary enforcement look politically loaded. This is why brinkmanship is so corrosive. It forces the state to choose between underreaction and overreaction while the operator remains inside a zone of plausible argument. In Arctic and sub-Arctic fisheries, where every boarding or escort can be read as signalling, that choice is unusually difficult.

The quiet motif throughout is the minute. A minute to answer the radio. A minute before gear is hauled. A minute while a patrol decides whether to close. A minute of delay that lets weather worsen or a boundary fall astern. Strategic brinkmanship thrives in these narrow slices of time because it does not need to defeat the state completely. It only needs to use the minute better. Fisheries governance, by contrast, is often slower, because it must think about evidence, precedent, safety, and the next season as well as the next move. That asymmetry is the tactic's true advantage.

The larger lesson is that incidents at sea should not be read only as immediate compliance events. They are also tests of institutional character. How much ambiguity will be tolerated? How much delay will be accepted as caution rather than obstruction? How quickly will a technical fisheries issue be allowed to turn into a political stand-off? Operators who practice strategic brinkmanship are not only chasing fish. They are collecting answers to those questions. Once collected, the answers circulate.

That is why the connection between micro-behaviour and macro-dispute dynamics is so important. A state may believe it is handling one awkward vessel. A fleet may understand that it is watching the state set the tone for the next five seasons. If a handful of minor incursions and slow-walked orders gradually establish that certain waters are effectively negotiable at sea, then the strategic environment of the fishery has already changed, even without a single formal legal ruling. Governance loses ground in increments long before it loses face in public.

The unresolved question, then, is not whether some operators will continue to test limits. They will. The harder question is what kind of enforcement posture can deny them useful precedent without converting every marginal incident into a full diplomatic performance - and whether authorities can do so consistently enough that the minute no longer belongs to the vessel that is boldest at using it.

Chapter 8

Ports, Markets, and Leverage

In 2009, the FAO adopted the Port State Measures Agreement, and with that step a quiet truth about fisheries governance moved closer to the centre of policy. A state did not need to catch every suspect vessel at sea in order to shape behaviour. It could wait at the quay. It could inspect papers, deny services, question origin, or refuse landing. The harbour gate mattered.

That shift deserves more attention than it usually receives. Fisheries politics is often narrated through patrols, boundaries, and confrontations on the water, as though control begins where nets enter the sea. Much of it now begins later and elsewhere. Fish become valuable only when they are landed, processed, financed, insured, certified, and sold. That means governance can work through the ordinary routines of commerce as much as through direct coercion. A vessel may outrun a patrol. It cannot so easily outrun the need to unload, document, and convert catch into money.

The central tension lies here: markets can reinforce public regulation, but they can also blunt, reroute, or quietly deform it. Port inspections, traceability systems, retailer standards, banking due diligence, and insurance requirements all appear to strengthen oversight by creating more points at which doubtful fish can be stopped or questioned. Often they do. Yet each additional gate also redistributes power. Larger operators may absorb compliance costs more easily than smaller ones. Trade may be redirected rather than deterred. Measures that look stringent in law may become selective in practice if demand remains resilient and alternative channels stay open. Governance widens. So does adaptation.

This matters because ports and markets solve a problem patrols cannot solve cheaply. Arctic and sub-Arctic fisheries unfold across immense spaces under hard weather, thin inspection capacity, and moving stock distributions. Sea-based enforcement remains essential, but it is rarely sufficient. Routine commerce offers leverage at lower marginal cost. A vessel denied landing, refuelling, transhipment services, or market access feels pressure differently from one merely observed at a distance. That pressure can be more predictable, more documentable, and often more scalable. It also changes the meaning of enforcement. The decisive question becomes not only who can stop a vessel, but who can deny it a legitimate route into value.

The stakes extend well beyond compliance in the narrow sense. Ports and markets decide which fish become visible, which records are believed, which firms remain bankable, and which communities bear the cost of tighter control. A traceability rule can reduce laundering and still push burdens down the chain towards smaller operators least able to manage them. A certification scheme can strengthen discipline and still privilege actors already integrated into the most documented export routes. A sanction can signal seriousness and still move trade into darker channels where states know less and control less. None of this makes market governance futile. It makes it political. Every gate disciplines some actors more than others.

That is why the chapter's real subject is leverage rather than simple regulation. Ports are leverage points because fish, documents, fuel, labour, and shipping schedules converge there. Processing hubs are leverage points because fleets dependent on a handful of cold-chain nodes do not negotiate from the same position as fleets with multiple outlets. Finance and insurance are leverage points because permission to operate is dispersed across institutions far from the fishing ground itself. Even reputational pressure, for all its fragility, can matter when operators need stable buyers and clean paper. The line between public and private governance begins to blur. Not disappear. Blur.

And yet leverage is rarely linear. A tighter rule does not always produce a cleaner fishery. Restriction may harden compliance in one corridor and increase opacity in another. A port denial may deter one operator and redirect another towards a more permissive jurisdiction. A retailer standard may reward transparency while quietly concentrating market power. A sanction may satisfy domestic political demand while reducing the very visibility on which future enforcement depends. These are not implementation mistakes alone. They are the ordinary consequences of governing through markets populated by actors who adapt quickly and unevenly.

The path ahead follows that adaptation across the key nodes where fisheries become governable after the catch. It begins at the port, where routine inspection and documentation can do work patrols cannot. It then moves through traceability and certification into the private ordering of access, before turning to the bargaining power created by processing geography and cold-chain dependence. From there the argument darkens through sanctions and bans that may redirect rather than stop trade, then widens into finance and insurance as gatekeepers of operational permission, before returning to the communities whose consent, resentment, or exclusion determines whether leverage remains politically sustainable.

The harder question, then, is not whether ports and markets can govern fisheries. They plainly can. The sharper question is what kind of leverage they create when fish are already moving through unequal supply chains, unequal states, and unequal communities - and whether a system that grows stronger at the quay may still weaken itself if the price of control is to make lawful trade more concentrated, opaque trade more inventive, and coastal legitimacy more fragile than before.

Port state measures as governance tools

In 2009, the Food and Agriculture Organization adopted the Agreement on Port State Measures to Prevent, Deter and Eliminate Illegal, Unreported and Unregulated Fishing. The achievement sounded administrative. It was more than that. The agreement recognised a governing fact that Arctic and sub-Arctic fisheries had been teaching for years: a state need not catch every suspect vessel at sea to shape behaviour. It can wait at the quay. It can inspect documents, deny entry, refuse landing, withhold services, and force a vessel to prove more than its presence. The gate can matter more than the chase.

That shift deserves to be taken seriously because it alters the entire logic of fisheries control. Patrols remain important, especially where deterrence depends on visible state presence. But sea-based enforcement is expensive, weather-bound, and unevenly available across northern waters. Port controls exploit a simpler truth. Fish are not valuable merely because they are caught. They become valuable when they are landed, serviced, documented, processed, and sold. A vessel may avoid inspection on the fishing ground. It cannot so easily avoid the need to turn catch into commerce. Routine trade becomes leverage.

This is why port state measures are best understood not as secondary paperwork but as strategic governance. The show is unglamorous: advance requests for entry, catch documentation, inspection of holds and papers, verification of licences, comparison of declarations with known records, refusal of services where serious doubts persist. The name for this is port-based control. The explanation is straightforward. Ports compress uncertainty. They gather vessel, cargo, paperwork, fuel needs, labour turnover, and commercial intent into one place where scrutiny is cheaper than on the open sea. The implication is powerful. A state with modest patrol reach can still change incentives significantly if vessels know that doubtful catch may become commercially useless at landing.

This matters especially in Arctic and sub-Arctic fisheries because distance magnifies the price of offshore certainty. A patrol aircraft may identify something suspicious without being able to prove enough to justify stronger action. A coast guard vessel may arrive too late, or under conditions that make boarding unsafe or politically fraught. Port inspection, by contrast, often occurs under calmer procedural conditions. The cargo is no longer moving in the same way. Documents can be checked against one another. Inspectors can compare declared species, quantity, area of capture, and authorisation status with more time and less weather risk. What could not be settled at sea may become intelligible ashore. The fish stop moving. That changes everything.

The European Union's long use of port and trade-based fisheries controls offers one clear example of this logic at work. When states or fleets are accused of undermining agreed fisheries management, the decisive pressure has often come not from dramatic interdiction but from the ability to restrict access to ports, landings, or related services. The dispute with the Faroe Islands in 2013 over herring and mackerel showed how market and port measures could become materially consequential without requiring overt confrontation at sea. The broader lesson was not simply that Europe had commercial weight. It was that denying a route into value can bite harder than trying to shadow every vessel on the grounds.

A second example lies in the broader implementation of port state measures after the 2009 agreement. The agreement did not imagine ports as a magical filter that would stop all doubtful fish. It recognised them as choke points. A vessel requesting entry must identify itself, describe its catch, declare its fishing authorisations, and expose itself to possible inspection. That requirement changes behaviour before any hold is opened. Operators who know their paperwork, identity chain, catch origin, or quota use may not survive scrutiny must now factor in a real possibility that the trip will fail commercially at the moment it should have become profitable. Port control works because it makes future denial part of present calculation.

This is the key to deterrence in the port context. Deterrence does not come only from punishment after a completed offence. It comes from making the expected value of doubtful conduct less attractive in advance. A vessel operating in a remote northern fishery may tolerate uncertain monitoring at sea if it believes that landing, refuelling, resupply, or transhipment services remain available afterwards. Remove that assumption and the whole voyage changes meaning. The abstraction here is incentive structure. Its visible form is a decision: whether to risk a dubious transhipment, whether to keep fishing near a contested seam, whether to file incomplete documentation, whether to take on catch whose origin cannot later be defended. Ports govern by changing those decisions before the vessel docks.

And yet one should not romanticise port measures as a frictionless solution. They work well only when documentation systems are credible, inspectors are trained, agencies share information, and legal authority is clear enough that refusals and detentions can survive challenge. A port officer who cannot compare declared catch against licence conditions or prior vessel behaviour is merely collecting paper. A state that threatens denial but rarely follows through will teach operators the wrong lesson. Port control is cheaper than constant offshore enforcement, but it is not cheap in the larger administrative sense. It requires databases, inspection protocols, legal discipline, and political support strong enough to withstand complaints from industry or foreign governments when a vessel is turned away.

This is why routine commerce matters so much. A port is not just an enforcement site. It is where fisheries administration meets shipping schedules, processor demands, labour changes, customs systems, and local economic pressure. The show can look deceptively ordinary: a request for berth, a cold store booking, a fuel order, a need for repairs, a consignment waiting for onward transport. The name is commercial dependency. The explanation is that vessels, even powerful ones, remain dependent on mundane services. The implication is that denying those services can shape conduct more efficiently than dramatic offshore pursuit. A trawler without a berth, a buyer, or a way to offload has not been physically seized. It has still been governed.

A useful limitation must be admitted. Port state measures can deter wrongdoing more cheaply than patrols, but they do not eliminate adaptation. Operators may redirect catches to more permissive ports, rely on transhipment, shift into less transparent trade routes, or use intermediaries that blur origin and ownership. A strict port can therefore improve its own compliance posture while pushing opacity outward. This is not a reason to abandon port-based governance. It is a reason to understand that leverage redistributes behaviour rather than freezing it. A denied landing in one harbour may become a hurried search for another. Control succeeds in one node and weakens in another.

That is why ports become strategic nodes rather than final solutions. They are points where the state can concentrate attention and force dispersed activity into administrative contact. They are not the whole fishery. A vessel that never asks for entry cannot be inspected in the same way. A catch transferred at sea, redirected through a different jurisdiction, or processed through less demanding channels may escape the intended scrutiny. The strength of port control lies not in universal reach but in selective pressure. It makes compliant channels more attractive for some actors and riskier for others. Governance through ports is therefore about shaping the map of movement, not sealing it entirely.

Port measures also have a political dimension that deserves emphasis. Refusing entry or services is often portrayed as technical enforcement, but it can have the feel of a sanction even when framed administratively. That makes ports powerful and sensitive at once. A state that uses port controls consistently and transparently can build credibility. A state that applies them selectively, or appears to use them as covert foreign policy in some cases and routine administration in others, may erode trust. Fisheries operators notice patterns quickly. So do governments. The difference between principled control and opportunistic exclusion lies not only in the law but in visible practice.

This is particularly important in northern fisheries where livelihoods and regional economies may depend heavily on a small number of landing points. Port measures that look elegant from a governance perspective can feel punitive or arbitrary to local communities if they disrupt work, depress throughput, or shift catch away from already fragile coastal economies. The administrative logic may be sound. The legitimacy effects may still be damaging. A harbour denied fish for good enforcement reasons can nevertheless become a site of political grievance if those reasons are not understood or if burdens fall unevenly. This is one reason port governance cannot be treated as a mere technical add-on. It enters directly into the politics of compliance.

The quiet motif throughout is the gate. A gate is ordinary. It is also decisive. At sea, authority is stretched by distance, weather, and uncertainty. At the gate, the vessel must ask. It must request entry, submit records, identify itself, and accept the possibility that commerce will stop where the patrol never arrived. That is why ports matter so much in fisheries governance. They convert routine dependence into leverage and routine paperwork into selective power.

The deeper lesson is that fisheries control often works best where it looks least dramatic. A denied berth, a document mismatch, a refusal of landing services, a request for further clarification before cargo can move: these do not resemble the classic imagery of enforcement. They may matter more. They force operators to calculate not only what they can get away with on the water, but what they can still monetise once the trip ends. For remote and thinly patrolled fisheries, that shift is profound. It changes governance from pursuit to conditional permission.

The unresolved question is therefore not whether port state measures can be effective. They plainly can. The harder question is how far a state can rely on the gate before adaptation, trade diversion, and local political resentment begin to thin the very legitimacy and visibility that make port-based leverage so powerful in the first place.

Traceability, certification, and retailer pressure

In 1997, Unilever and the World Wide Fund for Nature launched what became the Marine Stewardship Council, and a new kind of fisheries governance began to take visible institutional form. No coast guard was created. No maritime boundary was redrawn. Instead, pressure entered the fishery through labels, audits, chain-of-custody requirements, and the purchasing decisions of firms far from the grounds themselves. A fish could now be disciplined after capture by the need to remain legible in commerce. The label travelled.

That development matters because it changed the location of control. Fisheries had long been governed through law, patrols, port inspections, and inter-state bargaining. Certification and traceability added another layer: private actors began to demand proof not only that fish existed and could be sold, but that they could be traced, narrated, and defended through a chain of documents and standards. This did not replace public regulation. It complicated it. Markets ceased to be merely where the outcome of governance was realised. They became places where governance itself was performed, contested, and, at times, outsourced.

Traceability is the most straightforward version of this shift, though only in appearance. The show is familiar enough: a catch is landed, documented, moved through processing, packed, shipped, and sold under a chain-of-custody system meant to preserve knowledge of where it came from and under what authorisation. The name for this is commercial traceability. The explanation is simple. If fish can be followed reliably from capture to market, laundering becomes harder, dubious catches become riskier to handle, and buyers gain leverage over upstream conduct. The implication is substantial. A vessel that escapes notice at sea may still be disciplined later if it cannot insert its catch cleanly into a market that demands traceable origin.

This promise is real. So are its complications. Traceability systems do not merely reveal information; they impose a way of organising work. Fish must be segregated, documented, coded, audited, and moved through approved channels with sufficient consistency that later verification remains possible. Larger firms often absorb this more easily because they already possess compliance staff, digital systems, established buyer relationships, and a scale of operation that can distribute administrative costs. Smaller operators, remote processors, and less capitalised fleets often experience the same requirements differently. For them, traceability may feel less like governance through clarity and more like governance through paperwork. The form is the same. The burden is not.

This is why private standards can reinforce public regulation while also distorting it. A state may welcome certification and retailer demands because they create pressure against illegal, unreported, or weakly documented fish where direct enforcement remains thin. A buyer may insist on chain-of-custody integrity that public authorities cannot police continuously themselves. The result can be a genuine strengthening of the system. Yet the same structure may shift costs and exclusion downwards. Operators able to comply smoothly retain access to premium or more stable markets. Others, including some whose actual fishing practices are lawful and sustainable, may find themselves marginalised because they cannot satisfy the evidentiary and audit culture demanded by private schemes. Governance then begins to sort fishers by administrative capacity as much as by conduct.

The Marine Stewardship Council provides an instructive example because it shows both the attraction and the friction of certification as governance. Its standards and chain-of-custody requirements have helped create a commercial language through which buyers, processors, and retailers can reward some fisheries and discipline others. This has often reinforced public management where states and regional bodies were already trying to align conservation, documentation, and market access. But the same system has also generated disputes over cost, fairness, and accessibility. Smaller fisheries and producers have long argued that certification can privilege industrial actors better equipped to handle audits, consultants, documentation routines, and the slow administrative labour of proving what they do. The problem is not that standards exist. It is that standards are never socially neutral once they become gateways.

A second example lies in the European Union's use of catch certification and import controls as part of its broader fisheries governance. These measures seek to reduce the entry of illegal, unreported and unregulated fish into the European market by demanding documentary assurance about origin and legality. This is public regulation, not private certification in the strict sense, but it reveals the same political structure. Market access becomes conditional on traceability. A fish that cannot be documented convincingly loses value no matter how profitably it was caught. That can strengthen deterrence. It can also reallocate commercial opportunity towards actors with better administrative infrastructure and towards routes less likely to trigger costly scrutiny.

The deeper issue is that markets do not merely consume regulation. They interpret it. Retailers, processors, wholesalers, and brand owners all decide what level of evidentiary confidence they require before they are willing to handle a product. Their decisions can amplify state efforts or blunt them. A retailer that insists on strong chain-of-custody discipline and rejects suspect consignments becomes, in effect, a secondary enforcement site. A retailer that treats certification as a marketing ornament while tolerating weakly verified supply weakens the system, however polished its public language may be. The show is commercial. The name is private enforcement pressure. The explanation is that firms govern upstream conduct whenever their buying rules alter what kinds of fish can be sold profitably. The implication is that a supermarket specification may matter to a vessel captain more than a distant legal text the captain expects never to encounter directly.

This is why reputational sanctions attract such attention. Certification loss, public naming, buyer blacklisting, or media exposure can all injure firms and fisheries without requiring a patrol to board a vessel or a court to impose a penalty. Reputational pressure is attractive because it appears faster, cheaper, and more scalable than direct enforcement. In some markets, it is. A firm with strong brand exposure and dependence on quality-sensitive consumers may respond quickly to avoid contamination of its image. Yet reputation has limits, and those limits matter most where demand is resilient. If buyers exist who care less about origin, if alternative markets remain open, or if the product can be redirected through intermediaries that dilute the public association with wrongdoing, reputational sanctions lose force. A damaged name matters only where the market insists on remembering it.

That limitation is crucial because it prevents the romance of ethical consumption from outrunning institutional reality. Private standards and retailer pressure can reduce laundering and improve discipline, but they do not abolish the incentives to reroute product into darker channels. When enforcement or certification pressure becomes costly enough, some operators adapt by changing buyers, processors, documentation pathways, or jurisdictions. The fish still moves. It simply moves somewhere less visible. This is not a reason to dismiss traceability. It is a reason to understand that traceability works inside market geography. Where market demand is broad, fragmented, or relatively indifferent to provenance, reputational tools may discipline only part of the fishery while leaving another part surprisingly resilient.

A useful limitation should also be admitted on the other side. Not every cost imposed by traceability is unjust or merely bureaucratic. Documentation, audit, and segregation are often the price of making laundering and origin fraud harder. To complain that a fishery must prove what it sells is not, by itself, an argument against the system. Many of the burdens that smaller operators experience are the same burdens that make the system credible. The challenge is not to remove all cost. It is to decide whose costs matter, how they are distributed, and when the pursuit of traceability begins to produce a fishery that is cleaner on paper but narrower in participation and legitimacy.

This is why markets must be treated as arenas of governance rather than as passive endpoints. A fish entering a chain-of-custody regime is not merely entering commerce. It is entering a structure of conditional permission. Auditors, certifiers, buyers, retailers, and data systems all become part of what determines whether that fish will travel smoothly, be questioned, or be refused. This can be immensely valuable where public enforcement is thin. It can also generate a governance style that is indirect, uneven, and only partly accountable to democratic scrutiny. A state can be blamed for exclusion that a private buyer effectively imposes. A certifier can shape the fortunes of a fleet without carrying the political burden that a regulator would bear for a comparable decision. These arrangements are powerful precisely because they are dispersed.

The quiet motif throughout is the label. It looks simple, even banal: a small sign on a box, a package, a purchase order. Yet behind it sits a dense political economy of proof. The label condenses scientific advice, documentation discipline, audit practice, buyer preference, and reputational calculation into something that can move through a warehouse or a supermarket chilled aisle. It is a small object carrying a large claim. Whether that claim is trusted, affordable, and enforceable determines much of the real force of modern market governance in fisheries.

The consequences return, inevitably, to legitimacy. A traceability or certification regime that improves confidence while visibly excluding smaller or more remote operators can generate quiet resentment that later resurfaces as political resistance to the wider governance system. A retailer-led demand for impeccable chain-of-custody may delight regulators and brand managers while burdening the very communities whose support matters most for compliance and long-term adaptation. Legitimacy is therefore not a separate moral flourish added after the market has done its work. It is part of whether the market can continue to do that work without creating opposition strong enough to undermine it.

This leaves a harder question than advocates of private standards sometimes acknowledge. If traceability, certification, and retailer pressure make some laundering more difficult and some supply chains cleaner, but do so by shifting costs, concentrating market access, and leaving resilient demand available elsewhere, what exactly should count as success: a more documented fishery, a more exclusionary one, or a market system that can reduce opacity without slowly teaching smaller lawful operators that governance now belongs less to the state than to the label?

Processing hubs and value-chain dependence

In 2013, when the European Union moved against the Faroe Islands during the herring and mackerel dispute, the immediate argument was about fisheries management and market access. The deeper argument was about where power really sits once fish leave the net. A fleet can catch successfully and still remain vulnerable if only a small number of ports, factories, cold stores, and buyers can turn that catch into money. The fish may be free in the water. The value is not.

That distinction matters because fisheries governance is often described as though control lives mainly in licences, patrols, and quotas. Those things matter immensely. Yet a fishery is also a chain of dependencies. It needs berths, cranes, ice, grading lines, freezers, boxes, transport, customs paperwork, and onward buyers. In remote northern waters, it needs these even more acutely because distance punishes delay and product quality can deteriorate quickly once the catch is aboard. A fleet that depends on a small number of landing points does not bargain from the same position as one with multiple processors competing for supply. The harbour chooses back.

Processing hubs are powerful because they condense both economics and verification. The show is familiar: a vessel lands in one of only a few ports capable of handling the species, volume, and cold-chain demands of the trip. Inspectors, plant managers, customs officers, and buyers all become relevant at once. The name for this is nodal power. The explanation is simple. Where fish must pass through a narrow set of facilities to retain commercial value, whoever governs those facilities gains indirect influence over conduct at sea. The implication is significant. Enforcement feasibility depends not only on who can board a vessel offshore, but on who can credibly say no when the catch reaches its most vulnerable point in the value chain.

This is why processing capacity creates leverage even when it is not exercised in openly punitive ways. A plant can demand consistent documentation before accepting fish. A port can insist on inspection and delay unloading until records are reconciled. A state can make service access conditional on prior compliance. A buyer may prefer catches routed through facilities known for strong controls because those facilities reduce reputational and legal risk. None of this requires dramatic enforcement theatre. The fishery is governed through ordinary commercial dependency. A vessel that can evade direct scrutiny at sea may still behave differently if it knows that a small number of processing hubs stand between its catch and payment.

The North Atlantic provides repeated examples of this logic, even where formal disputes are narrated in terms of quotas or sanctions. The pressure exerted on the Faroe Islands in 2013 mattered because trade and port access were linked to a fishery already dependent on particular market and processing routes. The issue was never only who was right on herring and mackerel allocation. It was also who controlled the nodes through which disputed fish had to pass to become fully valuable. Once those nodes narrowed, bargaining power shifted. The lesson extends well beyond that dispute. A fleet is more governable when its options ashore are fewer.

Iceland offers a different kind of illustration. Its fisheries are often analysed through quota design and ownership concentration, rightly so, but the geography of processing has long mattered just as much. Cold stores, reduction plants, filleting facilities, export logistics, and specialised ports do not merely receive catch. They organise the economic meaning of access. A vessel that depends on a particular landing pattern or on processors able to handle a certain species mix is tied to infrastructure that can reward, punish, or redirect conduct indirectly. The fish is caught on the grounds. The bargaining position is often decided ashore.

This is where economics enters enforcement in a more intimate way than the language of "compliance" usually suggests. If a regulator or state wants to influence behaviour, it helps enormously to know whether a fleet can reroute catch easily or whether it is effectively captive to a small number of hubs. Captive is a strong word, but the underlying condition is common enough. Small or specialised fleets may have nowhere else to go without major losses in time, quality, or price. Remote northern operations are especially exposed because steaming to an alternative processor is not an abstract inconvenience. It can mean spoilage risk, missed market timing, crew fatigue, and fuel cost that turns a legal catch into a bad commercial decision. A fish kept cold keeps power.

That phrase captures the political economy of the cold chain. Chilling, freezing, storage, and onward transport are not neutral services. They determine whether a catch remains commercially viable long enough to travel, and they do so in ways that concentrate advantage in the hands of those who own or regulate scarce infrastructure. A state with well-placed processing capacity can exert influence over foreign or contested fleets without constant offshore confrontation. A firm controlling key cold-chain assets can demand stricter documentary assurance than the state itself can practically verify on the fishing ground. The show is logistical. The name is infrastructural leverage. The explanation is that product quality and market timing make some facilities more indispensable than legal texts alone would suggest. The implication is that economic geography can discipline behaviour where patrol reach is weak.

This also explains why states and firms that control key nodes sometimes become more powerful than their formal jurisdiction might imply. A country may not be able to police every seam at sea, but if its ports and plants are among the few realistic destinations for a profitable landing, it gains indirect regulatory reach. A processor may not write law, but if fleets need its lines running and its buyers engaged, it can insist on standards that shape practical compliance. This is not a conspiracy of private power replacing the state. It is a layered order in which enforcement becomes feasible partly because commercial infrastructure narrows choice.

Still, this leverage is not automatic. A processing hub can influence behaviour only if three conditions broadly hold. The fleet must need it, the facility must be willing to impose conditions, and alternatives must not be too easy to reach. Break any of those, and nodal power weakens. A fleet with multiple landing options can play one port against another. A processor dependent on volume may accept fish it privately distrusts. A state that fears economic disruption in a coastal town may hesitate to use a strategic node as forcefully as legal logic suggests. This is a useful limitation because it prevents a simplistic view in which infrastructure ownership automatically translates into discipline. Dependency is a relationship, not a guarantee.

Another limitation follows from adaptation. When governance tightens around a key hub, trade may be redirected rather than cleaned. Vessels may seek more permissive ports, rely more heavily on transhipment, or shift towards buyers willing to accept weaker traceability. A narrow and well-governed node can therefore push some activity into broader and murkier channels. This does not mean the hub has failed. It means value-chain leverage, like other forms of fisheries governance, alters behaviour unevenly. A stricter node may increase visibility for the compliant and opacity for the evasive at the same time.

And yet the significance of processing hubs remains because they change the cost structure of non-compliance. A vessel considering aggressive behaviour at sea has to ask not only whether it can avoid inspection offshore, but whether it can still land profitably into the infrastructure on which it depends. This is where economics and enforcement meet most directly. A doubtful haul is one thing. A doubtful haul with no credible processing route is another. Regulators understand this instinctively, which is why they often focus less on abstract legal symbolism and more on where fish must queue. They govern where the fleet cannot easily improvise.

This is particularly important in fisheries built around a small number of landing points. In such systems, the concentration that improves efficiency can also strengthen governance capacity - up to a point. A limited number of ports makes inspection and documentation easier to organise, creates institutional memory among inspectors and processors, and reduces the universe of commercial exits through which questionable catch can disappear. But the same concentration can create political fragility. If a community depends heavily on one plant or one hub, any restrictive action may carry heavy local costs. What is economically efficient for the supply chain may become socially combustible for the town. The very node that gives the state leverage may also make the state cautious.

This is why processing geography belongs inside any serious account of compliance. It shapes what enforcement can realistically do, how quickly it can bite, and how politically survivable it is when it does. A system with strong offshore rules but diffuse or weak processing control may look robust until product starts moving through the least scrutinised route. A system with modest patrol capacity but strong nodal control may outperform expectations because it governs where fish become money. Neither outcome is accidental. Both are products of where the value chain narrows.

The quiet motif throughout is the gate, though here the gate is cold, industrial, and humming rather than ceremonial. A vessel may range widely at sea. It often has far fewer options when the catch needs ice, grading, freezing, and a buyer before its value slips. That gate is where economics becomes a form of jurisdiction. It decides what can move onward and what must wait, explain itself, or lose value.

The deeper lesson is that fisheries enforcement is never only about catching wrongdoers. It is about understanding dependency. Which fleets can reroute and which cannot? Which species require specialised plants? Which ports can absorb volume without asking too many questions? Which cold-chain nodes are so central that their standards effectively shape behaviour far beyond the quay? Once those questions are asked seriously, the map of governance changes. The patrol route remains important. So does the processing line.

The unresolved question, then, is not whether processing hubs create bargaining power. They clearly do. The harder question is how that power should be used when the same concentrated infrastructure that makes compliance easier to demand also makes whole communities more vulnerable to disruption - and when pushing too hard at one node may simply send fish, information, and control towards another that is colder in all the wrong ways.

Sanctions, bans, and unintended consequences

In August 2013, the European Union imposed measures against the Faroe Islands during the dispute over Atlanto-Scandian herring and north-east Atlantic mackerel. The immediate intention was clear enough: increase the cost of behaviour judged incompatible with agreed fisheries management. The immediate effect was clear too: access narrowed, pressure rose, and the dispute moved from quota arithmetic into trade and port politics. Yet the deeper lesson was less comfortable. Restriction does not act on empty water. It acts on routes, buyers, processors, and governments that will adapt. Close one door and the fish do not stop moving.

That is the central problem with sanctions and bans in fisheries governance. They are often treated as the hard edge of seriousness, the point at which regulation proves it can bite. Sometimes they must bite. A state or regional body that never denies access, restricts trade, or raises the cost of defiance invites contempt. But tougher rules do not travel through the fishery in a straight line. They move through markets, paperwork, shipping decisions, and opportunistic intermediaries. A sanction rarely lands on fish alone. It lands on a network. The route changes first.

This matters because restrictive measures are designed to do two things at once: punish undesirable conduct and reshape future incentives. The show is familiar. A market is restricted, a landing is denied, a vessel is refused service, a consignment is questioned, a firm is cut off from preferred buyers. The name for this is coercive economic governance. The explanation is that many fisheries actors fear blocked commerce more than distant patrols, because commercial denial turns a completed trip into a costly failure. The implication appears promising. If the catch cannot be sold or serviced, the risk calculus at sea should change. And sometimes it does. But only sometimes. That is where the interesting and harder story begins.

The first difficulty is substitution. Fish are marketable goods, and marketable goods seek alternative channels when pressure rises in one corridor. A sanction directed at one jurisdiction, one port system, or one class of buyers may reduce volume there while encouraging rerouting through more permissive processors, less demanding traders, or states less interested in strict verification. This is not a side effect. It is a predictable adaptation. Operators, firms, and even governments will ask not only whether a route has closed, but where the product can still move with tolerable profit and tolerable scrutiny. The punishment succeeds in one sense. It may fail in another by making the fish harder to see.

The 2013 dispute with the Faroe Islands makes this point well because it showed both the force and the limits of restrictive measures. European pressure mattered because the relevant markets, ports, and commercial relationships mattered. The Faroes could not simply shrug. Yet the dispute also demonstrated that coercive measures do not end political contest by themselves. They harden bargaining, encourage search for alternative outlets, and recast the conflict as much in terms of recognition and fairness as in terms of conservation. A trade measure does not arrive as pure deterrence. It arrives as a story about who is being disciplined, by whom, and for what purpose. That story shapes how adaptation proceeds.

A second example comes from the broader use of import controls and trade restrictions against fish suspected of being illegal, unreported or unregulated. These measures can be effective where the targeted operators depend heavily on markets that value documented legality and where alternative buyers are scarce or commercially inferior. But effectiveness falls when demand remains resilient elsewhere. A product excluded from one market may reappear through another after processing, relabelling, transhipment, or sale into channels that require less documentation and fewer reputational assurances. The fish remains real. The knowledge thins. That thinning matters because governance needs information as well as pressure. A sanction that reduces visibility can weaken later control even while satisfying the immediate desire to act decisively.

This is why linear thinking is such a danger. It is tempting to imagine a simple ladder: more severe rule, more compliant behaviour, cleaner fishery. Fisheries markets rarely behave so obediently. Tighten an import regime and some operators improve documentation, while others redirect cargo, restructure ownership, or rely more heavily on intermediaries who can absorb uncertainty. Ban port services and some vessels avoid the risky conduct, while others shift towards at-sea transfer, remote landings, or less transparent harbours. Restrict insurance or finance and some firms reform, while others seek capital in places that ask fewer questions. The rule bites. The system bends.

The deeper reason is that sanctions act on incentives embedded in unequal supply chains. Larger firms with diversified buyers, access to legal advice, and the capacity to reorganise routes can often absorb restrictive measures better than smaller operators. States with multiple export corridors adapt differently from those dependent on a few buyers or hubs. This means sanctions are not only coercive instruments. They are sorting mechanisms. They distinguish between those who can reconfigure and those who cannot.

That can be useful, even necessary. It can also create perverse outcomes where the more traceable and visible actors bear the sharpest immediate cost, while the more opaque and opportunistic actors prove surprisingly resilient.

A useful limitation must be admitted. None of this means sanctions and bans are futile. In some circumstances they work precisely because adaptation options are limited. A fleet dependent on a narrow set of ports or premium buyers may change behaviour quickly when denied access. A processor whose business model depends on trusted certification may become far more cautious once reputational or legal pressure rises. A state facing concentrated market dependence may return to the negotiating table because the alternative is plainly worse. These are real effects. The argument is not against restrictive measures. It is against the fantasy that they operate in isolation from the adaptive intelligence of those targeted by them.

That is why governance must anticipate adaptation rather than merely react to it. A sanction should be read not as an endpoint but as the opening move in a second-round contest over route, ownership, product form, and market channel. What will the vessel do if denied this port? What will the trader do if this market closes? What documentary substitutions become more attractive? Which intermediaries gain by moving into the shadow created by the restriction? These are not implementation afterthoughts. They are part of the sanction itself. A rule that fails to anticipate rerouting may win a public battle and lose the informational war that follows.

This brings knowledge back to the centre. Restrictive measures can succeed politically because they are visible. They reassure domestic audiences that authorities are acting. They signal seriousness to partners and rivals. But fisheries governance cannot live on signal alone. It also depends on knowing where product goes, who handles it, and which channels remain open. If tougher measures push activity into routes with poorer traceability, weaker inspection, or thinner reporting, states may find themselves with less usable intelligence than before. The fishery may become formally stricter and practically murkier at the same time. Enforcement then grows blinder while sounding tougher.

A further complication is timing. Sanctions often deliver their sharpest effects later than political rhetoric suggests. A denied service, blocked landing, or restricted import can disrupt immediate plans, but operators learn quickly. Contracts are rewritten. New buyers are cultivated. Flags change. Carriers are rerouted. Processing moves. The adaptation may not be elegant. It need only be sufficient. This is one reason why early assessments of restrictive measures can be misleading. Immediate pain is visible. Longer-term redistribution of activity is less so. Governance that congratulates itself too soon may miss the moment when the fishery has already reorganised around a different and less transparent equilibrium.

This is where the motif of the route matters. Fisheries governance is always, in part, route governance. A catch becomes governable because it must pass through identifiable paths of landing, paperwork, finance, processing, and sale. Sanctions try to close some of those paths. The success of the measure depends not only on the closure itself but on the geometry of what remains open. A route closed in one jurisdiction can become a gift to another. A transparent channel made too costly can strengthen a murkier one. The map must be read dynamically, or not at all.

There is also a legitimacy question that cannot be ignored. Sanctions and bans may be legally justified and politically necessary, yet still appear selective or unfair to those who bear the cost. Coastal communities that lose throughput, crews whose lawful work is disrupted, or smaller firms unable to reroute as easily as larger rivals may interpret restrictive measures not as principled governance but as punishment distributed by market power. That perception matters because legitimacy feeds back into compliance. A regime seen as strategically intelligent and broadly fair can survive disruption better than one seen as blunt, hypocritical, or indifferent to uneven consequences.

The harder truth, then, is that restrictive measures are strongest when they are embedded in a broader strategy of tracing, co-ordination, and adaptation management. Bans without information may displace rather than deter. Sanctions without parallel work on documentation, port intelligence, and supply-chain comparison may narrow one channel while widening another. Toughness without foresight becomes a kind of administrative theatre - satisfying for a moment, then increasingly costly as control leaks into the less visible spaces the tougher rules have themselves helped create.

The unresolved question is not whether sanctions and bans should ever be used. They must be. The sharper question is what kind of restrictive measure can raise the cost of wrongdoing without merely teaching fish, firms, and states to travel by routes that are dimmer, less documented, and harder to govern than the ones just closed - and whether authorities are willing to judge success not by the drama of the restriction, but by the quality of control that remains once the route changes.

Insurance, finance, and operational permission

In 2016, when the European Union tightened its rules on the public record of fishing vessels, the administrative target was not only the ship on the water. It was the wider circle of actors asked to trust that ship: insurers, lenders, service providers, buyers, and ports. A vessel does not operate on fuel and steel alone. It operates on permission. That permission is rarely granted by the state alone.

This is the first point that matters. Fisheries governance is often imagined as a vertical relationship between regulator and operator: licence issued, rule broken, sanction imposed. The real structure is broader and less tidy. A vessel needs insurance to move credibly through commerce, finance to acquire and maintain gear, access to working capital to survive long seasons and volatile catches, and documentary legitimacy to reassure those who underwrite, transport, or buy the fish. These are not peripheral concerns. They are the conditions under which a trip becomes commercially plausible. A bank officer, an insurer, or a credit committee can shape conduct at sea without ever seeing the sea.

That makes insurance and finance powerful forms of indirect enforcement. The show is deceptively ordinary: a lender asks for ownership clarity, compliance history, and documentation before extending credit; an insurer demands information about vessel identity, flag, route, or regulatory standing before issuing cover; a renewal is delayed because a case remains unresolved; a premium reflects reputational and legal risk rather than only weather and machinery. The name for this is financial gatekeeping. The explanation is straightforward. Where risk is monetised, uncertainty about legality becomes a cost rather than a distant moral issue. The implication is substantial. A vessel that can evade inspection may still struggle if it cannot prove itself insurable, bankable, or commercially normal.

This matters especially in Arctic and sub-Arctic fisheries because operating costs are high, voyages are long, and interruptions are expensive. Distant fishing is capital-intensive even before one begins to discuss compliance. Cold-water vessels, specialised gear, fuel, crew turnover, maintenance, and cold-chain timing all require reliable access to capital and risk transfer. That dependence gives insurers and lenders leverage they do not always realise they possess. A bank that refuses to fund a vessel with opaque ownership or weak documentation is not merely protecting itself. It is helping define the boundary of who can fish at scale. An insurer that asks harder questions about flag changes, sanction exposure, or compliance history may be doing as much practical governance as a maritime patrol whose range is limited by weather and budget.

A useful example comes from the long struggle against illegal, unreported and unregulated fishing, where public authorities and non-governmental investigators alike have repeatedly found that vessel identity, ownership opacity, and shifting registration matter not only to regulators but to insurers, financiers, and the firms that rely on clean risk assessments. The vessel may change name, flag, or operator. The underlying commercial challenge remains the same. Someone must decide whether the risk is acceptable. In that moment, governance becomes dispersed. The decisive question is not only whether the state can punish. It is whether the vessel can still secure the practical permissions that make punishment less necessary in the first place.

A second example sits closer to market governance. Once catch certification and traceability schemes became more significant in international seafood trade, financiers and insurers gained new reasons to care about regulatory credibility. A vessel or company operating under weak documentation no longer posed only a legal risk. It posed a commercial continuity risk. Buyers might hesitate, ports might scrutinise more sharply, processors might refuse consignments, and sanctions could spread outward through the value chain. The bank or insurer need not become a fisheries policeman to respond to this. It need only treat poor compliance evidence as bad business. That is often enough to change terms, raise cost, or deny support altogether.

Still, one should not exaggerate the purity of this private leverage. Insurers and lenders are not guardians of the fishery by instinct. They are risk managers and profit-seekers. Their influence depends on transparency, on the quality of records available to them, and on whether the operator truly lacks alternatives. This last point is crucial. Financial gatekeeping works best where access to capital and insurance is concentrated in institutions that care about compliance, reputation, and legal exposure. It weakens when operators can move into alternative credit, opaque ownership vehicles, state-backed arrangements insulated from outside pressure, or insurers willing to accept murkier portfolios for a price. Financial discipline is therefore powerful but uneven. It bites hardest where markets are already relatively visible.

That limitation keeps the analysis honest. It prevents a familiar fantasy in which private finance appears as a clean substitute for public enforcement. It is nothing of the kind. A vessel denied cover by one insurer may seek another. A company refused ordinary bank finance may rely on internal capital, politically protected lenders, or intermediaries less sensitive to scrutiny. Some operators are especially good at arbitraging exactly these differences. They do not escape permission entirely. They shop for it elsewhere. The state may then find that pressure applied through respectable financial channels has pushed the problem into darker and less documented forms of capital where knowledge declines along with control.

This is why transparency matters so much. Financial and insurance pressure works only when the relevant institutions can identify whom they are dealing with, under what ownership structure, with what flag history, in what regulatory standing, and with what commercial exposure. If vessel identity is blurred, beneficial ownership hidden, sanction history fragmented across jurisdictions, and compliance data inaccessible or untrusted, the lender or insurer cannot gatekeep effectively even if it wishes to. The weakness is not moral. It is informational. The same opacity that frustrates fisheries enforcement frustrates the wider institutions that might otherwise reinforce it. A signature is only as strong as the file behind it.

The file returns here as a quiet motif because it captures the administrative reality of dispersed permission. A vessel may appear powerful on the water, yet its operating life depends on files in offices far away: underwriting files, loan files, due diligence files, company records, claims histories, sanction checks, beneficial ownership declarations. None of these alone determines conduct. Together they decide whether a trip is routine, suspicious, overpriced, or impossible. This is the larger point. Permission to operate is not a singular act. It is a chain of tolerances.

That chain can shape behaviour before any breach occurs. A company planning to reflag opportunistically, operate with weak documentation, or use logistics that complicate catch origin may decide against it if the resulting uncertainty threatens financing or insurance on unacceptable terms. The deterrent effect here is real, though often invisible. Unlike a patrol stop or a denied port entry, it may leave no dramatic public trace. A trip is modified before departure. A vessel is sold, restructured, or idled. A lender insists on a cleaner ownership arrangement. An insurer excludes certain routes, practices, or counterparties. These are not side effects. They are governance through anticipated commercial friction.

The problem, however, is that such friction can be distributed unevenly. Smaller operators with thin margins and less administrative support may feel financial scrutiny more sharply than larger firms equipped with compliance departments and legal teams. A private gatekeeping regime that looks prudent from a risk perspective may therefore deepen concentration within the sector. Larger companies can document more, hire specialists, and reassure counterparties. Smaller firms may face higher premiums, worse credit terms, or slower access to working capital even where their actual conduct is not more problematic. Governance through finance can therefore improve discipline while also narrowing participation. It does not simply reward virtue. It rewards administrability.

This is one reason the relationship between private gatekeeping and public legitimacy remains delicate. If fishers and coastal communities experience financial compliance demands as arbitrary, remote, or skewed towards large integrated firms, they may read the whole governance system as one in which access to the fishery depends less on good fishing than on proximity to powerful institutions and clean paperwork. That perception matters even when the underlying risk logic is not wholly wrong. Legitimacy in fisheries is never just about ecological outcomes. It is about whether the rules of commercial survival appear reasonably shared. A system that outsources effective permission to operate to banks and insurers cannot ignore how unequal that permission may look from the harbour.

A useful counterpoint should be admitted. In some cases, financial and insurance scrutiny can support smaller lawful operators by making opacity and identity games more expensive for larger and more aggressive rivals. A vessel owner who fishes carefully, maintains clean records, and operates transparently may benefit when lenders and insurers begin asking harder questions across the sector. This is the most persuasive version of financial governance: not private power substituting for the state, but private caution helping make lawful practice more competitive. Yet even this version depends on whether institutions can distinguish genuinely lower risk from merely better-resourced presentation. That is never automatic.

The broader lesson is that enforcement must be understood as a dispersed ecology rather than as a single sovereign act. The state still matters decisively. Licences, inspections, sanctions, and legal rules set the formal frame. But the practical ability to fish depends on many smaller grants of confidence issued elsewhere. Insurance says the voyage is coverable. Finance says the operator is solvent and acceptable. Buyers say the product is saleable. Ports say the vessel is serviceable. Certification bodies and auditors may say the chain is credible. None of these institutions should be idealised. All can fail, discriminate badly, or be circumvented. Yet together they help explain why some vessels remain ordinary commercial actors and others drift towards the margins where opacity is both strategy and necessity.

This dispersed permission also changes how one should think about sanctions. A formal sanction may matter less than the way it is interpreted by insurers, banks, and counterparties. A vessel not yet banned may still become commercially toxic. The reverse is also true. A vessel under suspicion may continue operating if enough financial and commercial actors remain willing to tolerate uncertainty. The state does not govern alone, but neither do private institutions. Governance emerges from how these judgments accumulate and whether they converge. When they do, the operator feels pressure long before a final legal ruling. When they diverge, ambiguity lives.

The unresolved question, then, is not whether insurers and lenders can act as gatekeepers. They clearly can. The harder question is what kind of fisheries regime makes their gatekeeping useful rather than distortive: one in which transparency is strong enough that financial caution reinforces public regulation, but not so unevenly distributed that permission to operate becomes, in practice, a privilege for the already legible and already powerful while more opaque capital quietly welcomes the rest.

Community impacts and legitimacy feedback

In August 2013, when the European Union imposed measures against the Faroe Islands during the herring and mackerel dispute, the argument was presented in the language of orderly management and market discipline. The legal and institutional stakes were real. So were the community effects. In a small fisheries-dependent society, a restriction on market access does not remain a technical correction for long. It becomes a judgement felt in harbours, plants, council chambers, and households. The dock remembers quickly.

That is where legitimacy begins to matter in a harder sense than policy language often allows. Port controls, certification requirements, trade restrictions, and financing conditions may all be designed to improve compliance or deter conduct judged harmful to the fishery. Many do exactly that. But coastal communities do not encounter these measures as abstract governance tools. They encounter them as interruptions to work, as uncertainty over landing schedules, as reduced throughput for processors, as thinner margins for vessel owners, and as signals about whose costs are considered acceptable. A rule may be technically sound and still politically destabilising if those living closest to the fishery experience it as remote discipline imposed through markets they do not control.

This is why community impacts should not be treated as sentimental afterthoughts to otherwise rational regulation. The show is visible enough: a landing is denied, a processor loses volume, a port handles fewer vessels, a buyer demands paperwork smaller operators struggle to provide, a credit line tightens, a seasonal workforce shrinks, local representatives begin to speak less about compliance and more about unfairness. The name for the resulting problem is legitimacy erosion. The explanation is simple. A governance measure that imposes concentrated local pain while distributing benefits more diffusely invites a political reading that may overwhelm its technical rationale. The implication is serious. Compliance begins to look less like collective stewardship and more like externally managed sacrifice.

This matters especially in Arctic and sub-Arctic fisheries because the economic and social geography of the coast is often narrow and exposed. A small number of landing points, processing plants, service providers, and seasonal labour channels may hold together the practical life of an entire district. Remove or destabilise one part of that chain and the effect is not spread evenly across a national economy. It lands in a place. That concentration sharpens political response. A minister may see improved traceability or stronger deterrence. A coastal town may see an idle line, a postponed trip, or the quiet disappearance of younger workers who no longer believe the fishery offers a future. Governance that does not account for such concentration often mistakes local anger for mere resistance to reform, when it is more often resistance to being made the buffer for adjustments designed elsewhere.

A useful example lies in Iceland's long disputes over quota ownership, transferability, and the distribution of fisheries value. These arguments are usually framed around efficiency, resource rent, and concentration, but they also illuminate the feedback loop between market structure and public legitimacy. When communities believe that rights, landings, or processing decisions are being detached from place and reorganised chiefly for commercial efficiency, support for the wider management regime can become conditional and brittle. The issue is not that regulation becomes useless. It is that people begin to ask whether the system still recognises their dependence as politically meaningful. That question travels far beyond quota design. It shapes how later port restrictions, traceability demands, and market exclusions are interpreted.

The Faroe Islands dispute offers a second example from the market side rather than the ownership side. External restrictions may have been intended to alter state behaviour in an international management conflict, but their domestic meaning was inseparable from livelihoods and political recognition. In such settings, a measure aimed at improving governance can deepen local conviction that governance is being exercised against the community rather than with it. This is not simply a communication failure. It reflects the structure of cost distribution. When a policy's immediate burden is local and visible, while its longer-term benefits are collective, disputed, or externally framed, legitimacy becomes harder to maintain. The pressure then feeds back into national negotiating positions, electoral politics, and tolerance for future compromise.

This feedback loop is the crucial point. Legitimacy is not a moral ornament added after enforcement has done its work. It is part of whether enforcement remains governable over time. A community that broadly accepts the fairness of a difficult measure may still suffer under it, but it is more likely to continue co-operating with inspections, documentation demands, and politically inconvenient conservation decisions. A community that experiences the same measure as arbitrary, selective, or blind to local dependence will often respond differently. Compliance becomes thinner, public support for regulators weakens, and politicians begin to treat technical adjustments as existential tests of loyalty. The fishery does not simply become harder to regulate. It becomes harder to describe in common terms.

This is why port and market measures can generate outcomes that look contradictory from a narrow policy perspective. A rule may reduce laundering, improve documentation, or strengthen buyers' confidence, while simultaneously eroding trust among fishers, processors, and local governments expected to bear its heaviest burdens. The abstraction here is policy effectiveness. Its visible counterpart is whether people still co-operate when the next rule arrives. If they do, legitimacy has held. If they comply minimally, contest everything, or begin to support evasive behaviour as a justified defence against unfair pressure, then a technically effective measure may already be undermining the conditions of its own repetition.

A reasonable limitation should be admitted. Community grievance is not always evidence that a measure is unjust. Some rules hurt because they close off genuinely harmful practices, expose long-tolerated opacity, or force economic restructuring that was avoided too long. A fishing town can sincerely feel wronged and still be defending arrangements that were unsustainable, exclusionary, or politically protected beyond reason. Legitimacy does not mean that every local objection is correct. It means that governance cannot remain stable if those objections are ignored as though social acceptance were optional. The hard task is to distinguish between pain that accompanies necessary discipline and pain that signals a politically dangerous mismatch between policy design and lived coastal dependence.

This distinction becomes sharper once markets are recognised as arenas of governance rather than neutral endpoints. Certification schemes, retailer standards, insurance requirements, and documentation rules may all push the fishery towards greater visibility. They may also privilege larger operators and better-capitalised supply chains that can absorb audit costs, administrative burdens, and temporary disruption. Smaller ports and firms often feel this asymmetry first. They may lose throughput not because their fishing is less legitimate, but because their administrative capacity or bargaining power is weaker. Market governance then appears less like principled regulation than like a sorting mechanism that rewards scale and paperwork fluency. That perception matters whether or not it is fully fair.

The political consequences can spread well beyond the immediate measure. Coastal representatives begin to speak in the language of abandonment. Regulators become easier to caricature as tools of distant ministries, foreign markets, or large firms. Industry organisations may harden against reforms they might otherwise have accepted. Even scientifically grounded restrictions can be reinterpreted through this accumulating distrust. A denied landing or lost contract today becomes evidence, tomorrow, that the whole regulatory order is indifferent to local continuity. Once that reading takes hold, every future intervention enters a poisoned field. Legitimacy does not disappear all at once. It thins through repetition.

This is why feedback is the right metaphor. Market measures affect community livelihoods. Community experience shapes political support. Political support influences how governments enforce, negotiate, and communicate subsequent measures. Those subsequent choices then alter how markets behave in relation to the fishery. The cycle can be constructive. A port regime seen as fair may strengthen documentation culture, improve buyer confidence, and stabilise local expectations. It can also become corrosive. A scheme seen as punitive may encourage resistance, evasive adaptation, or populist pressure for looser rules even where ecological and commercial discipline remain badly needed. The same technical instrument can therefore produce opposite long-term governance effects depending on how its burdens and benefits are socially distributed.

The quiet motif here is the dock. Policy arrives there last and is judged there first. A denied berth, a slower unloading process, a new documentary demand, a vanished buyer, a season with fewer landings: these are the practical forms in which international agreements, certification regimes, and enforcement strategies become real to coastal communities. The dock is where abstraction turns into memory. That memory shapes the next encounter between the state, the market, and the fishery.

The broader lesson is that legitimacy must be understood as infrastructure. Not softer than inspection, not secondary to markets, but part of the practical system that allows difficult rules to be repeated without each one becoming a fresh political rupture. A government that ignores this may still enforce effectively for a while. It will govern on thinning ice. Once coastal support erodes far enough, even technically sound measures begin to trigger counterproductive politics, and the fishery becomes harder to stabilise through law or commerce alone.

The unresolved question, then, is not whether port and market measures can work. They can. The harder question is what makes them feel fair enough to those whose jobs, ports, and local futures are bound up with the fishery that stronger control is meant to protect - and what happens when the people asked to carry the discipline begin to believe that the market is being governed more carefully than the community.

Chapter 9

Enforcement Capacity in Distant Seas

A vessel appears on a screen near a seam of jurisdiction, then slows in a way that suggests more than transit. The weather is turning. The nearest patrol asset is occupied elsewhere. The legal basis for inspection exists, but exercising it will cost time, fuel, paperwork, and perhaps a diplomatic explanation by morning. Nothing in this scene is dramatic by itself. That is precisely the point. Fisheries enforcement in distant seas is rarely defeated by a single grand failure. It is thinned by ordinary choices made under pressure, scarcity, and uncertainty.

The popular image of enforcement is simple: a rule is broken, a state responds, order is restored. The working reality is more exacting. A patrol vessel may be available on paper and irrelevant in practice if another mission already has priority. A boarding team may be trained and still unable to act safely in the conditions at hand. A suspicious event may be detected and still never ripen into a case if the evidence chain begins badly or the administrative machinery behind it is too slow. Capacity, in other words, is not the same thing as authority. A hull is not a verdict.

That distinction matters because distant-sea fisheries expose the gap between what states can legally claim and what they can operationally sustain. The sea is large, the weather punishing, the missions competing, and the opportunities fleeting. Under those conditions, fleets learn quickly what is actually policed rather than what is merely prohibited. They observe which waters are watched closely, which incursions draw only radio warnings, which infractions survive into prosecution, and which disappear into the fog of overloaded agencies and unfinished files. Enforcement becomes intelligible to operators not through doctrine, but through pattern.

The central tension lies here: states need enforcement strong enough to shape behaviour, yet every serious act of enforcement in remote waters carries cost, risk, and political meaning. A boarding is never just a boarding.

It is a judgement about priorities, safety, evidence, and escalation. A prosecution is never just a sanction. It is the endpoint of a long chain that can fail at many points for reasons that have nothing to do with innocence. This is why distant-sea enforcement should be understood less as a matter of will than as a pipeline under strain. The patrol at sea and the file ashore belong to the same problem.

The stakes are larger than compliance in the narrow sense. Under-enforcement does not merely allow some rule-breaking to continue. It teaches a fleet how the system breathes. If operators come to believe that harsh conditions, budget pressure, or institutional overload will usually blunt follow-through, then strategic behaviour begins to look rational rather than reckless. A vessel edges closer to a line. A response is delayed. A case collapses. The lesson travels. Over time the fishery itself adjusts to the true, thinner boundary of risk. Law remains printed. Practice moves.

And yet the answer is not simply more force. Distant enforcement can harden incidents rather than settle them, especially when boarding rules, evidentiary standards, and jurisdictional sensitivities are already contested. Authorities know this. So do operators. That mutual awareness creates a dangerous middle ground in which both sides calculate not only legality but interpretation: how an inspection will look, how a refusal will sound, how a shaky case may affect the next season, the next negotiation, the next co-operative arrangement. The result is a field in which prudence and hesitation are often indistinguishable until after the moment has passed.

This is why institutions matter so much. Ships and aircraft extend reach, but they do not by themselves create credible enforcement. Credibility depends on budgets, mission priorities, boarding protocols, evidentiary discipline, prosecutorial stamina, and, crucially, the co-operative habits that allow states to share burden without multiplying friction. Joint patrols, working relationships, and information exchanges are not decorative additions to hard power. They are often what prevents routine fisheries control from becoming a series of isolated and politically loaded encounters. When those habits weaken, capacity shrinks even if the hardware remains.

The path ahead follows this pressure where it is felt most sharply. It begins with assets, money, and competing priorities, where the real shape of enforcement emerges long before a vessel is ever approached. It then moves into the friction of boarding itself, where safety and law meet the politics of presence, before turning ashore to the documentary and administrative pipeline that decides whether detection becomes consequence. From there the argument widens into co-operation, joint operations, and information sharing, where reach can be extended but never made immune to wider political weather.

The harder question, then, is not whether distant seas can be policed. They can, sometimes well. The sharper question is what counts as genuine capacity when every stage of enforcement is vulnerable to substitution, delay, risk, and institutional fatigue: the patrol asset on the horizon, the boarding rule that may never be used, the file that must still survive ashore, or the co-operative routines that keep the whole system from learning the wrong lesson about how much non-compliance it can quietly bear.

Patrol assets, budgets, and priorities

In January 2019, when the Russian trawler *Melkart-5* became the focus of a dispute around Svalbard, the visible question concerned authority, inspection, and response. Beneath that immediate episode sat a less dramatic but more durable problem. A state may possess the legal power to act, yet still face the harder choice of whether to commit scarce assets, in difficult conditions, to one fisheries incident rather than to search and rescue, sovereignty patrol, environmental response, or another task already competing for the same hull and crew. The vessel on the horizon is only half the story.

That is the right place to begin because distant-sea enforcement is often discussed as though law were the main constraint. It is not. Law decides what may be done. Resources decide what is likely to be done. A patrol ship can only be in one place at once. An aircraft launched for fisheries monitoring cannot simultaneously serve another mission. Crews tire. Maintenance schedules intrude. Weather narrows windows. Fuel, time, and political attention all carry cost. Capacity is therefore not an abstract national attribute. It is a sequence of choices made under pressure. The patrol plan reveals the truth.

This is why enforcement should be treated as a resource allocation problem as much as a legal one. The show is familiar: a suspicious vessel track appears in a remote zone; commanders compare distance, sea state, mission priority, crew readiness, and the probability that an intervention will yield usable evidence. The name for this is operational triage. The explanation is simple. States do not enforce all lawful possibilities equally because they cannot. The implication is uncomfortable but essential. What is actually policed reflects not only legal principle, but budget, distance, timing, and the opportunity cost of choosing one encounter over another.

Harsh conditions magnify every part of this calculation. Arctic and sub-Arctic waters are not merely far away. They are environments in which ordinary tasks become expensive and dangerous. Boarding in rough seas demands skill, time, and the willingness to accept physical risk. Long transits reduce time on station. Darkness, icing, and variable communications complicate already difficult work. A fisheries agency may write a rule assuming inspection can occur as needed. The sea does not read the rule. This is where rhetoric about strong enforcement often outruns reality. A state may own capable vessels and still face days or weeks in which effective fisheries policing is practically thin because the conditions make everything slower, costlier, and less certain.

Budget enters here not only as a matter of quantity, but of structure. A government can spend heavily on patrol assets and still underperform if maintenance, staffing, aviation support, and data analysis are mismatched. A single vessel unavailable for repair at the wrong time can reshape the real enforcement map more than an ambitious policy speech ever admits. An aircraft without sufficient crew rotation, a patrol ship without specialist fisheries boarding support, or a coast guard budget tied to wider political priorities all produce the same effect: legal authority remains broad while usable presence narrows. The difference between stated and lived capacity is often administrative before it is strategic.

Norway provides a useful example precisely because it is widely seen as comparatively capable in northern fisheries control. The Norwegian Coast Guard, fisheries administration, and associated monitoring systems form one of the stronger enforcement environments in Arctic-adjacent waters. That reputation is deserved. Yet even here, capacity is not infinite and never purely fisheries-specific. Patrol vessels and aircraft support multiple missions. Svalbard and adjacent waters are politically sensitive. Boarding and escort decisions can acquire diplomatic significance beyond ordinary compliance. A relatively strong state must therefore ration attention and weigh consequence. This is not a criticism of Norway. It is a reminder that if an enforcement-heavy system still works through prioritisation, weaker or more fragmented systems will do so even more sharply.

A second example lies in the Central Arctic Ocean agreement reached in 2018. The restraint built into that agreement made sense not only because ecological knowledge was incomplete, but because any future enforcement environment there would be exceptionally costly to sustain through direct presence alone. Parties understood, even without saying it crudely, that the farther north and farther offshore the fishery moved, the more unrealistic it became to imagine that legality would be guaranteed by ships and aircraft constantly watching. Prevention was cheaper than pursuit. That insight belongs squarely to the politics of resources. States accepted limits in part because they knew what full operational enforcement would require.

This helps explain why fleets behave strategically in ways that follow the enforcement map more closely than the legal map. Fishers and firms notice quickly which zones attract regular patrol attention, which seasons coincide with stronger presence, which seam areas are rarely visited except during particular operations, and which kinds of suspected conduct rarely produce follow-through. They do not need perfect information. They need only a working sense of where the state is likely to spend its scarce effort. Under-enforcement then becomes legible not as a mystery, but as a pattern. Once that pattern becomes predictable, it invites tactical adaptation.

That adaptation need not look like blatant lawbreaking. A vessel may fish more aggressively near a boundary known to be thinly patrolled. Another may treat reporting rules more casually during periods when inspection pressure is likely to be directed elsewhere. A fleet segment may exploit short, intense seasonal windows because it knows those windows overwhelm the state's practical capacity to be everywhere important at once. The name for this is strategic response to scarcity. The explanation is that operators learn from enforcement rhythms. The implication is that capacity gaps do not merely weaken deterrence in the abstract. They actively help structure where risk-tolerant behaviour becomes worthwhile.

A useful limitation must be admitted. Sparse enforcement does not always indicate failure or neglect. States make rational choices. A fisheries incident may genuinely rank below a distress call, an environmental threat, or another security task. Some waters may be governed acceptably through documentation, remote sensing, and port leverage rather than through constant patrol. A realistic assessment of capacity should not confuse selectivity with collapse. The problem begins when selectivity hardens into predictability that operators can read more accurately than the public can. Then scarcity stops being a neutral management fact and starts becoming part of the fishery's incentive structure.

This is why measuring capacity rhetorically is so dangerous. Governments like to cite ships, aircraft, budgets, and missions to demonstrate seriousness. Those indicators matter, but they do not answer the practical question that fleets care about: what happens when a suspicious event occurs in difficult conditions at the wrong time, in the wrong place, while other tasks already occupy the nearest asset? True capacity is revealed less by procurement announcements than by coverage patterns, deployment choices, and the frequency with which anomalies actually lead to inspection, evidence, and consequence. The patrol plan tells the truth more honestly than the press conference.

Competing missions deepen the difficulty because they make fisheries enforcement politically contingent. Coast guards and naval services rarely exist solely to police fish. They search for missing vessels, respond to pollution, assert sovereignty, assist scientific activity, and support wider security functions. In a tense political moment, fisheries may receive more attention because they symbolise control. In a different moment, they may slide down the list because another mission claims urgency. Fleets understand this interplay, even if imperfectly. A season of strong fisheries visibility can therefore be followed by periods of looser practical scrutiny not because the rules changed, but because the mission queue did.

This is one reason why under-enforcement is often misread. Commentators may treat a thin response as evidence of weakness in will, when the immediate cause lies in resource conflict, timing, or operational prudence. Yet motives matter less to the fleet than outcomes do. If the result is that certain activities are rarely challenged, the strategic lesson remains the same regardless of the state's internal reasoning. Capacity should therefore be judged from the operator's perspective as well as the ministry's. What matters is not only what the state intended to prioritise, but what pattern of practical permissiveness the fleet can infer from repeated choices.

The quiet motif throughout is the clock. A patrol asset steaming north is already spending political choice in hours. A suspicious vessel knows that an hour can mean the difference between boarding and no boarding, between a completed haul and a partial one, between weather that allows intervention and weather that closes the option. Budget pressures are ultimately translated into time: fewer days on station, longer maintenance gaps, slower response, less tolerance for speculative pursuits. Fisheries governance in distant seas is therefore governed partly by time converted from money and then spent against uncertainty.

This is why resource realism matters so much. Enforcement capacity should not be praised because it exists in law or procurement tables. It should be tested against the actual friction of distance, weather, competing missions, and institutional stamina. A state that admits those limits can build smarter combinations of remote sensing, port controls, seasonal focus, and co-operation. A state that denies them risks teaching fleets that rhetoric is strongest exactly where presence is thinnest. Under those conditions, the fishery begins to adapt to the gap.

The harder question, then, is not whether distant fisheries can be policed with enough ships, aircraft, and money. No system ever has enough of all three. The sharper question is what kind of prioritisation makes scarcity less exploitable by fleets that learn quickly where opportunity begins: the most symbolic deployments, the most legally obvious cases, or the quieter and more disciplined use of limited assets in ways that make the patrol map harder to predict than the budget would suggest.

Rules of boarding and operational risk

In January 2019, the Norwegian Coast Guard boarded the Russian trawler *Melkart-5* in waters around Svalbard, and a fisheries inspection rapidly became something larger: a test of authority, restraint, seamanship, and political nerve. The legal questions were serious enough, but the more revealing fact lay in the practice itself. A boarding is never only an administrative act, not in distant seas. It is a close physical encounter between institutions that may distrust each other, conducted in cold water, under time pressure, with consequences that can travel far beyond the vessel in question.

That is why boarding sits at the tense centre of fisheries enforcement. From shore, it can look routine - a lawful inspection, a check of papers, a look in the hold, a few questions about gear and catch. On the water, the same act acquires friction almost immediately. Weather matters. Speed matters. The relative movement of hulls matters. So does the crew's mood, the captain's judgement, the boarding team's training, and the political meaning already attached to the flag on the stern. A rope ladder is a small thing. It changes the argument.

The attraction of boarding is easy to understand. The show is direct: officers step aboard, inspect documents, verify identities, examine catch, check holds, compare logbooks with the vessel's actual condition, inspect gear, and ask questions that remote systems cannot ask with the same force. The name for this is direct verification. The explanation is that some facts can be tested properly only in physical proximity, where documents, fish, compartments, equipment, and personnel can be compared in real time. The implication is obvious. Boarding remains the clearest available bridge between suspicion and firsthand evidence. It is the moment when fisheries law stops being abstract and becomes tactile.

And yet that tactile quality is precisely what makes boarding risky. The first risk is physical. Boarding teams move between vessels in conditions that can shift quickly from difficult to dangerous. Ice, swell, darkness, wind, deck movement, and fatigue all matter. A captain may legally permit a boarding while still arguing that the chosen moment is unsafe. The enforcement authority may suspect delay. The vessel may genuinely fear injury. Both can be true at once. This is not incidental. Safety is not a polite preface to enforcement. It is one of the main reasons boarding decisions are always narrower in practice than law on paper suggests.

The second risk is evidentiary. A boarding may produce information, but it also creates a chain of interpretation that must survive later scrutiny. What exactly was seen, by whom, under what conditions, recorded how, secured where, and connected to which legal threshold? A hold may contain fish inconsistent with expectation, but quantity, species identification, time of catch, and relation to the vessel's declarations may still require painstaking reconstruction. A document may appear irregular, yet turn out to be poorly organised rather than fraudulent. A boarding officer may reasonably infer one thing while a lawyer later insists on another. The name for this problem is evidentiary fragility. The explanation is that physical access does not abolish ambiguity; it merely changes its form. The implication is that even successful boarding does not guarantee successful enforcement.

A third risk is diplomatic. This matters most in Arctic and sub-Arctic waters, where fisheries policing often takes place near disputed or politically loaded spaces. A boarding that one side calls routine can be read by another as selective assertion, unnecessary provocation, or the disguised rehearsal of a broader jurisdictional claim. The 2019 *Melkart-5* incident matters because it captured exactly this slippage. What began as fisheries control did not stay confined to fisheries control. Boarding had turned into signalling, or at least into something close enough that both parties were forced to calculate its meaning beyond the immediate inspection. This is commoner than public debate admits. The boarding party does not merely step onto a deck. It steps into a narrative already waiting to interpret the move.

This is why authorities in distant seas often prefer remote verification and port controls where possible. That preference is not necessarily weakness or bureaucratic timidity. It is operational prudence. Remote systems can identify anomalies, confirm location patterns, compare declarations with tracks, and flag behaviour that deserves further attention. Port inspections, in turn, allow fish, documents, and vessel identity to be examined under calmer conditions, with more time, more staff, and less immediate physical risk. The show is less dramatic: a vessel is told to proceed to port, documents are checked ashore, services may be conditioned, and the chain of evidence can be handled with greater administrative control. The name for this approach is deferred verification. The explanation is simple. Some states would rather inspect slowly and securely later than confront quickly and ambiguously now. The implication is important. Enforcement method is not just a technical choice. It determines how likely an incident is to harden into confrontation.

A useful example sits in the broad logic of port state control after the adoption of the Port State Measures Agreement in 2009. The agreement did not abolish boarding. It reflected a clear institutional judgement that fish can often be governed more cheaply, safely, and documentably at the moment of landing than at the moment of suspicion on the water. This was not a retreat from enforcement. It was a recognition that boarding is one instrument among several, and often not the one best suited to every case. A vessel that can be denied landing, inspected in port, or compelled to reconcile its paperwork under controlled conditions may be more effectively governed than one intercepted in a high-friction encounter that yields contested facts and elevated political heat.

Still, port preference has limits. Some evidence disappears if one waits. Gear can be altered. Fish can be mixed, sorted, or transferred. The sequence of activity that mattered at sea may become harder to reconstruct ashore. Boarding therefore remains indispensable for certain kinds of cases, especially where immediate physical conditions or suspected conduct make later verification inadequate. This is why the real problem is not whether boarding should be replaced. It is how states decide when its unique evidentiary value outweighs its operational and diplomatic costs. That decision is never purely legal. It is strategic in the fullest sense.

Strategic brinkmanship by operators makes this harder. A captain who understands the state's aversion to escalation may comply slowly, contest procedures, invoke safety, or insist on legal clarifications just long enough to force hesitation. Such behaviour does not have to amount to outright refusal to be effective. It may only need to raise the political and operational cost of the boarding by a few degrees. The boarding team then confronts an old problem in a new form. Should it press harder and risk converting a fisheries inspection into a larger incident, or ease off and risk teaching the fleet that ordinary resistance is worthwhile? Boarding is high friction partly because it invites exactly this kind of calculation.

A second example lies in the wider practice of joint fisheries enforcement in northern waters, including long-standing co-operative habits between states that share stocks and patrol burdens. Joint operations and shared procedures can reduce the heat of boarding because they normalise expectations. A vessel approached under a familiar regime with established communications and inspection routines is less likely to interpret the encounter as exceptional political theatre. This is one of the underappreciated virtues of co-operative enforcement. It lowers the symbolic temperature of physical contact. The same ladder can mean less when the process around it is familiar, professional, and insulated, as far as possible, from wider diplomatic suspense.

A credible limitation should be admitted. Not every contested boarding reflects brinkmanship or geopolitical signalling. Some difficulties are simply what difficult seas produce. Weather turns. Communication falters. Captains make prudent safety decisions that look evasive to a patrol crew already under pressure. Boarding officers may overread hesitation because they know too well how delay can be used strategically. This matters because the category of operational risk must remain honest. If every awkward encounter is treated as disguised provocation, enforcement becomes theatrical and trust collapses. The strength of boarding as a practice depends partly on disciplined interpretation.

That said, method always shapes escalation pathways. A state that relies heavily on at-sea boarding accepts a certain exposure to physical risk and political interpretation in exchange for immediacy. A state that prefers remote verification and port control accepts a different exposure: delayed evidence, possible loss of immediate leverage, and the chance that the crucial event will be altered before arrival. Neither path is pure. Each organises danger differently. The serious analytical mistake is to imagine that enforcement intensity can be measured simply by how often a state boards vessels. Sometimes restraint in boarding reflects weakness. Sometimes it reflects a rational judgement that the better place to govern is later, colder, and with more paper on the table.

This is why boarding rules matter so much. They do not merely set safety protocols or legal formalities. They define how an enforcement authority moves from observation to contact, what thresholds must be met, what communications are required, what kinds of non-co-operation change the situation, and how far discretion runs before a routine inspection becomes a politically sensitive operation. Rules, in this sense, are not only technical safeguards. They are escalation architecture. They decide whether a case remains inspectorial, becomes coercive, or is redirected into another enforcement path altogether. A badly designed boarding regime may produce either unnecessary confrontation or paralysing caution. A disciplined one can narrow the zone in which confusion becomes strategy.

The quiet motif throughout is the ladder. It is practical, unheroic, and unforgiving. It marks the point at which remote suspicion becomes embodied authority. One person climbs. Another waits above. The sea moves under both. Around that small structure gathers the full weight of fisheries law, operational safety, evidentiary hope, and diplomatic risk. No wonder states do not use it lightly. No wonder some operators learn to exploit the fact.

The deeper lesson is that enforcement capacity cannot be assessed by counting patrol assets alone. One must ask what those assets are willing and able to do when the moment of contact arrives. A state may possess ships and aircraft, but if boarding is too risky, too politically sensitive, or too evidentially fragile under prevailing conditions, then real capacity lies elsewhere - in remote monitoring, in documentary pressure, in port control, in co-operative habits that lower temperature before the ladder is ever lowered. Boarding remains central because it reveals the limits of all these alternatives. It is the method that shows most clearly how much enforcement depends on judgement.

The unresolved question, then, is not whether boarding is sometimes necessary. It plainly is. The harder question is what kind of enforcement system uses boarding sparingly enough to avoid needless escalation, but credibly enough that vessels do not learn the wrong lesson from the state's caution - namely, that physical contact has become so costly, politically loaded, or operationally awkward that ordinary resistance at sea is now part of the fishery's working grammar.

Evidence, prosecutions, and administrative follow-through

In January 2019, the dispute over the Russian trawler *Melkart-5* in waters around Svalbard drew attention because the encounter was visible, dramatic, and politically charged. Yet the most revealing question in any such episode begins after the patrol vessel turns away. What survives the incident once it becomes paper? That is where fisheries enforcement usually hardens into consequence or drains into memory. The report decides later.

This is the neglected centre of enforcement in distant seas. Detection is exciting because it looks like action: the track appears, the boarding occurs, the discrepancy is noticed, the cargo raises questions, the crew is spoken to, the state seems present. But no fishery is governed by detection alone. Anomalies must be recorded, evidence preserved, identities verified, legal thresholds matched, files transmitted, agencies aligned, penalties chosen, and decisions defended against challenge. Between observation and sanction lies a long administrative corridor. Most enforcement weakness lives there.

That is why fisheries control should be understood as a full pipeline rather than a moment of interdiction. The show is familiar enough. A surveillance system flags a vessel, officers inspect or review its conduct, a possible breach is identified, and the case enters administrative or judicial channels. The name for the next problem is follow-through. The explanation is simple. A case that is visible on the water may still be unusable in law if records are incomplete, the chain of custody is weak, the relevant regulation is badly drafted, or the wrong agency receives the file too late. The implication is severe. States can look vigilant while remaining unconvincing.

This matters especially in Arctic and sub-Arctic fisheries because distance already makes first contact difficult. If a rare detection opportunity is then wasted through poor documentation or inconsistent follow-up, the loss is greater than in more densely policed settings. A boarding in rough weather may produce hurried notes. A port inspection may rely on papers assembled across several jurisdictions. A surveillance anomaly may need to be linked to licensing records, vessel identity, and catch documentation under time pressure. None of this is unusual. It is routine. The problem is that routine complexity can make apparently strong cases collapse later, sometimes for reasons invisible to the public and instructive to the fleet.

A useful example lies in the broader enforcement difficulties that followed the European Union's landing obligation once it became fully applicable to quota-managed catches in 2019. The rule's public meaning was straightforward enough: fish that once might have been discarded should be landed and counted. Yet its actual enforceability turned on a much messier question. Could agencies document what happened at sea in a form sturdy enough to survive administrative and legal scrutiny? Camera footage, observer notes, logbooks, landing declarations, and quota accounts all had to line up. Where they did not, the issue was not merely non-compliance by fishers. It was evidentiary fragility inside the state. A rule can be clear. A case can still fail.

A second example comes from the long struggle against illegal, unreported and unregulated fishing, where authorities and regional bodies have often possessed enough information to suspect wrongdoing without possessing enough documentary coherence to impose durable sanctions. Vessel identity changes, uncertain beneficial ownership, fragmented catch records, and weakly co-ordinated flag-state responses can leave even well-founded suspicions stranded in procedure. This is not a marginal issue. It defines how deterrence is actually felt. Operators do not need to believe they are invisible. They only need to believe that visibility rarely matures into a penalty they cannot explain away.

The chain of documentation is therefore central. A photograph, a hold inspection, a disputed logbook entry, a radar trace, or a witness statement does not carry equal weight merely because it exists. It must be dated, attributed, preserved, cross-referenced, and interpreted through the relevant rule. Who saw the fish? Under what conditions? How were species identified? Was the catch weighed, estimated, or described by eye? What was said by the captain, and was it recorded contemporaneously? Did the officer handling the evidence know what threshold the later prosecutor or administrative body would require? The name for this is evidentiary architecture. The explanation is that facts do not become usable on their own. The implication is obvious and too often ignored. A vessel can be caught. The case can still leak.

This is why inconsistent penalties are so corrosive. A sanction regime that is severe in theory but erratic in practice teaches the fleet the wrong lesson. One operator may be fined lightly for conduct another treats as commercially worthwhile. A third may face delay so long that the case loses urgency or public attention. A fourth may be penalised heavily only because its paperwork was easier to organise rather than because its conduct was worse. These disparities need not arise from bad faith. They can result from overloaded agencies, different local legal cultures, weak drafting, or prosecutors who simply do not rank fisheries offences highly against other pressures. The practical result is the same. Enforcement becomes negotiable in the mind of the regulated sector.

Courts and administrative bodies matter here more than maritime commentary usually admits. The patrol vessel and the inspector are visible symbols of state presence. The clerk, case officer, tribunal, prosecutor, or administrative reviewer are not. Yet these quieter institutions decide whether evidence is sufficient, whether penalties are proportionate, whether procedural rights were respected, and whether a file moves quickly enough to matter. This is where "paper capacity" enters as something more than a metaphor. It means trained staff, coherent forms, interoperable records, legal continuity, and the mundane ability to move a case through institutional steps without letting it die from bureaucratic delay. A desk can sink deterrence as surely as a storm can.

The problem is often aggravated by fragmentation. Fisheries detection may sit with one agency, licensing with another, scientific interpretation with a third, prosecution with a fourth, and diplomatic implications with a fifth. Each may be competent on its own terms. Together they may still fail if no one owns the whole pipeline. A boarding report written for immediate operational needs may not answer the later legal question. A prosecutor may inherit a technically dense file with little practical context. An analyst may know the vessel's behavioural history but lack a route to insert that history into the formal case. Fragmentation does not always produce collapse. It does, however, make collapse easier and responsibility harder to trace.

A sensible limitation should be admitted. Not every dropped or weakened case indicates institutional failure in the crude sense. Some suspicions are genuinely inconclusive. Some boardings produce ambiguous facts. Some signals that look suspicious from afar turn out to have innocent explanations once records are reconciled. Due process matters. A fisheries regime that sanctions on poor evidence will damage its own legitimacy and may quickly become vulnerable to legal challenge. This matters because the answer to weak follow-through is not a crude hostility to procedural care. The answer is stronger documentation and better administrative design, not lower standards of proof.

That limitation sharpens the core point rather than softening it. The serious problem is not that states sometimes fail to punish. It is that they often fail in predictable and repeated ways that fleets can learn. If cases are delayed because forms are incomplete, if certain offences rarely survive into sanction because species identification is weak, if courts treat fisheries matters as minor, if overloaded agencies settle rather than pursue, then the regulated sector begins to map enforcement through the probabilities of administrative exhaustion. This is how under-enforcement becomes strategic terrain. The law stays in place. The pipeline becomes the true regulatory boundary.

The pipeline metaphor matters because it changes how capacity should be measured. A state with patrol vessels, aircraft, and monitoring systems may look formidable. If files collapse, penalties vary wildly, and agencies cannot sustain cases from observation to outcome, the visible front end of enforcement overstates the real one. Conversely, a state with modest offshore reach but strong documentation discipline, integrated case management, and credible sanctions may govern more effectively than its hardware suggests. The question is not only whether the state can see the breach. It is whether it can keep hold of it long enough to make the sighting matter. The report still decides later.

This also explains why fleets care so much about paperwork even when public debate romanticises action at sea. Paper is where ambiguity can be cultivated, challenge prepared, and momentum slowed. A clever operator may not need to defeat the boarding officer. It may need only to ensure that the officer's record leaves room for doubt, that the relevant company identity is one step removed, or that the documentation trail becomes cumbersome enough for the case to age badly. Paper capacity, from the state's side, is the answer to this. It means building institutions that treat reports, evidentiary continuity, and sanction consistency as core elements of maritime enforcement rather than as aftercare.

The quiet motif throughout is the report. It is easy to underrate because it looks like aftermath. In reality it is where the incident becomes governable or not. A hurried note, an unclear species identification, a missing signature, a broken chain of custody, a confused reference to the wrong licensing condition: these are small failures with large consequences. They explain why the difference between a symbolic inspection and a deterrent case is often less the drama of the encounter than the administrative quality of what follows it.

The broader lesson is unromantic and important. Fisheries enforcement in distant seas is never finished on the water. The sea produces suspicion, contact, and sometimes proof. Institutions ashore decide whether those things become a durable fact in law. A state that neglects this will tend to mistake visibility for capacity. It will invest in seeing more and wonder why the fleet still behaves as though much of what is seen remains survivable. The answer usually lies in the pipeline.

The unresolved question, then, is not whether detection matters. It does. The harder question is what kind of enforcement system can carry a difficult case from observation to sanction without losing clarity, momentum, or proportionality along the way - and whether governments are willing to treat the quiet competence of forms, files, reviewers, and consistent penalties as part of maritime strength rather than as its unglamorous clerical shadow.

Co-operation, joint patrols, and information sharing

In 1976, Norway and the Soviet Union established the Joint Norwegian-Soviet Fisheries Commission, later the Joint Norwegian-Russian Fisheries Commission, to manage shared stocks in the Barents Sea. The visible achievement was institutional rather than dramatic. No great chase ended. No single confrontation settled the issue. Instead, a standing practice took shape: shared scientific work, recurring negotiations, technical co-ordination, and the quiet exchange of information that allowed difficult waters to remain governable across years of wider strain. The radio stayed open.

That detail matters because distant-sea enforcement is rarely defeated by law alone or rescued by hardware alone. Ships, aircraft, and satellite systems extend reach, but they do not remove duplication, mistrust, or the wasted effort that follows when neighbouring states chase the same signal separately or interpret the same incident through incompatible routines. Co-operation alters this more efficiently than rhetoric suggests. A patrol does not become twice as present because there are two flags in the water. It becomes more effective when information, expectations, and follow-through are aligned tightly enough that one sighting, one inspection, or one anomaly can travel across institutions without being lost or needlessly repeated.

This is the first reason joint patrols and information sharing matter. The show is simple enough: one authority spots unusual activity, another holds the licensing data, a third has recent inspection history, and together they produce a clearer picture than any one of them could alone. The name for this is pooled reach. The explanation is practical. States operating in remote fisheries rarely possess enough assets to police all relevant waters continuously, and they waste scarce capacity if they repeat one another's work blindly. The implication is immediate. Co-operation is not a diplomatic accessory to enforcement. It is one of the few ways to make thin capacity behave like something thicker.

The Barents Sea remains the strongest example because it shows that routine co-operation can do more than share burden. It can normalise restraint. Where states patrol, inspect, and exchange information through established channels, incidents arrive in an environment already shaped by precedent and habit. A boarding is less likely to be misread as an improvised political gesture if it occurs within a known framework of contact. A suspicious track is less likely to become an argument about motive if recent information has already been exchanged. This is the quieter achievement of joint work. It lowers the temperature before the incident occurs. The patrol vessel is not only carrying officers. It is carrying expectations built ashore.

Those expectations matter because enforcement in distant seas is full of moments that could harden into something larger. A radio query can sound like routine verification or a challenge. A boarding can read as normal compliance activity or as selective assertion. A delayed reply can look like operational caution or strategic obstruction. Joint operations and shared rules do not abolish these ambiguities, but they can narrow them. The more familiar the procedures, the less room there is for each side to dramatise the encounter for later advantage. The rule becomes ordinary. That ordinariness is one of the most valuable security goods fisheries institutions can produce.

Information sharing carries the same logic into the administrative chain. A vessel's behaviour rarely makes full sense from a single dataset. Position records, boarding history, quota use, licensing status, ownership information, port inspections, and prior warnings all contribute to the pattern. Where those records remain trapped in separate national systems, enforcement becomes slower and more tentative. Where they move through stable co-operative channels, authorities can act with more confidence and less duplication. The name here is informational continuity. The explanation is that cases weaken when the relevant facts are scattered or delayed. The implication is that co-operation extends beyond the moment of patrol. It shapes whether observation becomes action or simply another note in another file.

A second example lies in the 2018 Agreement to Prevent Unregulated High Seas Fisheries in the Central Arctic Ocean. The agreement was preventative, but its institutional design reflected the same governing instinct. Faced with a region where direct enforcement would be costly and ecologically uncertain, the parties placed unusual weight on consultation, scientific exchange, and continuing co-operative structure. This was not a sentimental choice. It was an acknowledgement that distant-sea order often depends less on dramatic interception than on building enough shared practice that a future dispute will have somewhere to go besides unilateral action. The agreement's force rested partly in the idea that co-operation itself could delay, and perhaps prevent, the point at which raw enforcement would be tested.

Still, co-operation should not be romanticised. It has costs and limits. Information shared too slowly is often as useless as information withheld. Joint patrols can become bureaucratic performances if they are not tied to real follow-through. States may agree to reduce duplication while quietly protecting national discretion at the moments that matter most. A sensible limitation must therefore be admitted. Co-operation does not replace state interest. It organises it. Where interests diverge sharply enough, the same channels that once lowered temperature can become sites of passive resistance, procedural delay, or selective disclosure. The mechanism survives. Its spirit thins.

That fragility is especially visible when wider political relations sour. Fisheries co-operation is often praised for surviving broader diplomatic strain, and sometimes it does. Yet survival should not be confused with immunity. A joint working group may continue to meet while trust inside it weakens. Data may still be exchanged, but with more hesitation, less candour, and greater anxiety about how admissions will be used. A patrol encounter once handled through quiet contact may suddenly carry extra symbolic weight because the larger relationship has altered the interpretive frame around it. This is why fisheries institutions are best understood as shock absorbers rather than as shelters outside politics. They cushion pressure. They cannot abolish it.

That limitation does not weaken the case for co-operation. It sharpens it. The question is not whether joint mechanisms can create perfect alignment. They cannot. The question is whether they produce enough shared routine that enforcement remains possible under friction rather than collapsing into duplicated effort, public accusation, or cautious inaction. In distant seas, that threshold matters enormously. A state with modest assets but strong co-operative channels may govern more effectively than a better-equipped state working largely alone. Shared expectations, after all, are a form of reach. They tell fleets what kinds of behaviour will likely be noticed, compared, and acted on beyond the first national boundary they cross.

This is why institutions deserve to be treated as part of operational capacity rather than as something softer and separate. A radio protocol, a shared inspection format, a routine exchange between analysts, a standing point of contact for urgent queries, a known pattern of joint operations: these look modest beside patrol hulls and aircraft, but they do work those assets cannot do alone. They reduce duplication, quicken interpretation, and calm incidents by making them legible in a familiar frame. The state remains sovereign. The system becomes less lonely.

The quiet motif throughout is the radio. Not because technology itself is the point, but because the radio symbolises what co-operation changes. A vessel is sighted. A question is asked. A neighbouring authority is informed. A prior inspection is recalled. A joint understanding is activated. The sea remains cold, distant, and politically sensitive. Yet the encounter is no longer beginning from silence. That is often the difference between a fisheries problem and a wider incident.

The deeper lesson is that co-operation extends enforcement capacity not simply by adding more eyes, but by changing the meaning of what those eyes see. A suspicious movement becomes more actionable when shared records exist. A boarding becomes less inflammatory when joint expectations surround it. An overloaded patrol system becomes more intelligent when others carry parts of the burden. None of this is automatic. All of it depends on maintenance, trust, and the willingness to keep administrative channels alive even when national moods favour hardness over patience.

The unresolved question, then, is not whether co-operation can help. It plainly can. The harder question is what happens when the very mechanisms that extend reach and cool incidents become politically fragile at the moment they are needed most - and whether distant-sea enforcement can remain disciplined once the radio is still working, but the trust behind it has begun to fail.

Chapter 10

Disputes Before they Become Crises

A fisheries crisis rarely begins with a crisis mood. It begins with a radio call, a disputed position, a boarding request, a delay long enough to look intentional, a photograph circulated before the paperwork has caught up. Someone says the vessel crossed. Someone else says the line was unclear, the order excessive, the gear lawfully deployed, the inspection improperly handled. By the time ministers speak, the quarrel has already lived several lives: on the bridge, in an operations room, in a ministry inbox, in a newsroom, and in the imagination of coastal communities that hear insult more quickly than they hear technical nuance. The message arrives first.

That is why fisheries disputes need to be understood as pathways rather than shocks. What later appears as a sudden rupture is usually the visible crest of a sequence already in motion. A minor enforcement choice acquires symbolic weight. An ordinary detention becomes a test of dignity. A contested inspection hardens because each side begins to speak not only to the other, but to domestic audiences waiting to judge whether weakness has been shown. The event at sea matters. So does the pace at which meaning gathers around it.

This is the central tension. States need enough firmness to protect rules, crews, and bargaining positions, yet every visible act of enforcement can also become part of a larger political story that narrows room for compromise. Delay may calm a situation or invite more probing. Clarity may stabilise expectations or lock in precedent. Ambiguity may preserve diplomatic space or encourage risk-taking by those who hear hesitation in every careful word. The difficulty lies not in choosing between order and flexibility, but in recognising that each instrument of de-escalation can also become an instrument of escalation if used too late, too publicly, or without a credible procedural path behind it.

The stakes are high because fisheries disputes are peculiarly good at generating mismatched tempos. Fish move by season. Political outrage moves by image. Legal processes move by file. Domestic expectations move by rumour and repetition. A government may need time to verify positions, consult agencies, and compare records. A detained crew or seized gear can become a national symbol long before that work is finished. Once the public frame hardens, technical accuracy alone may no longer be enough to restore room for manoeuvre. Governance then becomes a race between documentation and narrative, and narrative often leaves port first.

This matters later as much as now. A dispute handled badly does not simply damage one season. It teaches fleets, ministries, and local constituencies what kinds of behaviour are likely to be tolerated, what kinds of ambiguity can be exploited, and how quickly ordinary enforcement can be translated into political grievance. A state that cannot create credible off-ramps teaches others to test limits more boldly. A state that publicises too aggressively may satisfy domestic pressure while poisoning the channels it will need in the next incident. A state that says too little may preserve formal flexibility while allowing others to define the meaning of events. These are not communication errors sitting outside governance. They are governance.

What follows turns on that practical truth. The route into crisis often begins in recognisable incident patterns and then runs through the procedural devices that can still cool matters if they are built early enough and used credibly enough. It passes through the strategic uses of ambiguity, where silence, escort, delay, and publicity all carry signals whether intended or not. It then enters the harder terrain where media, nationalism, and narrative compression can overtake fact, before reaching the legal and diplomatic forums that promise settlement yet often deliver only slower, more formalised forms of contest.

A useful caution belongs here. Not every fisheries dispute should be solved quickly, and not every ambiguity is a policy failure. Sometimes vagueness preserves a working relationship that a premature search for final clarity would damage. Sometimes an interim arrangement is more stabilising than a definitive ruling that arrives too late for the season that triggered the quarrel. Sometimes a hotline matters more than a legal triumph because it keeps the next boarding from being read as a national insult. Managed uncertainty is not elegant, but elegance is not the point. The point is whether the fishery remains governable while disagreement persists.

And yet managed uncertainty has a cost. If parties begin to treat every grey area as a tactical asset, then ambiguity ceases to be a cushion and becomes an invitation. If every inspection can be spun, every escort treated as a provocation, every interim measure mocked as weakness, then the institutions designed to absorb friction begin instead to store it. This is why shock absorbers matter so much in fisheries politics. They do not prevent impact. They decide whether impact is transmitted directly into public crisis or slowed long enough for procedure to do its work. When they fail, even routine incidents acquire the feel of constitutional moments.

The harder question, then, is not whether a crisis can be recognised once it has arrived. By then the answer is often obvious. The sharper question is what should count, earlier, as the real warning sign: the first alleged crossing, the first contested inspection, the first public image that outruns the file, or the first moment when all sides stop treating the incident as something to document and begin treating it as something they must perform.

Typical incident pathways

In January 2019, the Russian trawler *Melkart-5* became the centre of a dispute in waters around Svalbard after a fisheries inspection widened into a much larger argument about jurisdiction, authority, and conduct at sea. What mattered was not only the event itself, but the speed with which it acquired new meanings. A boarding became a legal claim. A legal claim became a political signal. A technical matter turned, step by step, into a dispute watched by ministries, media, and domestic audiences well beyond the vessel involved. The sequence mattered more than the spark.

That is the right way to think about fisheries incidents in distant seas. They rarely arrive as pure surprises. More often they follow recognisable pathways in which an initially narrow point of friction accumulates narrative, institutional, and political weight. A vessel is said to have crossed a line. A detention is framed as routine by one side and provocative by another. Gear is seized under ordinary enforcement logic, then recast as an insult to livelihood or sovereignty. The first facts do not disappear. They are quickly joined by interpretation. A small file opens, and several larger ones begin to gather around it.

The central claim is therefore simple but easy to overlook. Crises are usually processes, not shocks. The show is familiar: a boundary allegation, an inspection dispute, a gear dispute, a warning ignored or answered slowly, a vessel escorted rather than merely queried, a photograph or statement released before the documentation is fully reconciled. The name for this is pathway escalation. The explanation is that each step creates a new audience and a new stake. The implication is practical and severe. If one knows the pathway, one can often see the pressure building before public language admits that anything serious is happening.

Boundary allegations are often the first and most durable point of ignition because boundaries promise clarity that the sea rarely delivers. A track suggests incursion. A captain disputes interpretation. Drift, timing, gear position, signal reliability, and charted lines all become arguable at once. None of this means the allegation is frivolous. It means that boundaries generate exactly the kind of factual dispute from which a larger political story can be built. The line is visible. Its meaning is not yet settled. That unsettledness is enough to start mobilisation ashore. A local industry group hears threat. A ministry hears challenge. A coastal public hears disrespect. The vessel's actual position remains important, but it is no longer the only thing at stake.

Detention raises the temperature further because it makes state power visible and personal. Once a vessel is held, ordered to port, or physically constrained in some way, the incident acquires human faces and immediate narrative force. Crew welfare, delay, lost catch, disrupted schedules, and images of officers aboard ship can all outrun the technical case that justified the action. The name for this is embodied escalation. The explanation is straightforward. People react more strongly to detention than to abstract legal disagreement because detention looks like a decision already made. The implication is that even a lawful and measured detention can be read by outside audiences as punishment before the file is complete. That perception then narrows room for later quiet adjustment.

Gear seizure works differently but often no less powerfully. A net, line, or other fishing gear appears mundane until it is taken away. Then it becomes proof, loss, and insult at once. To an enforcement authority, seized gear may be an ordinary evidentiary measure or a necessary response to suspected non-compliance. To the operator and those watching from shore, it may signify humiliation, economic harm, and a judgment about guilt that seems to precede due process. The show is material: gear on deck, tags attached, photographs taken, work interrupted. The name is symbolic confiscation. The explanation is that gear is not only equipment. It is the means of livelihood made visible. The implication is that seizures often travel politically faster than inspectors anticipate, especially where fisheries are already bound up with identity and national feeling.

Contested inspections are especially fertile ground for escalation because they sit in the grey zone between routine and confrontation. An authority treats the inspection as standard procedure. The vessel contests timing, basis, method, or manner. Compliance becomes slow, partial, argumentative, or highly legalistic. What might have remained a technical interaction now begins to produce a second-order dispute about the inspection itself. Was it selective? Was it safe? Was it properly authorised? Did the vessel delay prudently or obstruct strategically? The important point is that the original fisheries issue no longer stands alone. A new controversy has been created around process. This often doubles the number of claims that must later be reconciled and gives political actors more material with which to harden public positions.

The *Melkart-5* incident remains useful because it showed how quickly these elements can layer. Inspection, refusal, escort, and protest did not function as isolated acts. They formed a chain in which each step altered the meaning of the next. By the time the matter was discussed publicly, it was no longer possible to speak only of fisheries enforcement. A second example lies in the wider history of "Cod Wars" confrontations between Iceland and the United Kingdom in the mid-twentieth century. Whatever the considerable differences in legal context and political scale, the underlying lesson endures: disputes over fisheries rarely erupt from nowhere. They gather around repeated encounters, contested enforcement acts, and accumulating interpretations until a pattern that once seemed manageable becomes difficult to separate from national prestige.

A useful limitation should be admitted. Not every boundary allegation, detention, seizure, or contested inspection becomes a serious dispute. Many are contained because the facts are relatively clear, the procedures are familiar, the parties are used to one another, and quiet channels remain trusted enough to absorb the friction. This matters because it prevents a melodramatic reading in which every fisheries encounter is treated as the prelude to crisis. The argument is not that escalation is inevitable. It is that escalation follows patterns, and those patterns become easier to recognise once one stops treating incidents as isolated sparks.

This is where domestic audiences enter decisively. A fisheries incident does not move only between vessel and state. It moves between the vessel and its home industry, between the enforcement authority and its own ministries, between both sides and media environments already primed for stories of intrusion, weakness, or overreach. Once those audiences attach themselves to the event, the incentives shift. A ministry may become less willing to compromise for fear of appearing soft. A fishing association may frame a routine inspection as an existential threat. A coastal community may interpret delay as indifference or escalation as courage. Narrative stakes then begin to shape the file itself. Facts no longer travel alone.

The recurring mistake in public commentary is to imagine that the crisis begins at the moment everyone notices it. By then the sequence is usually well advanced. The more serious work lies earlier: identifying the predictable escalation points before they harden. Did the first allegation involve a boundary, where factual disagreement and symbolic meaning combine easily? Did detention or seizure make the event legible in dramatic human terms? Did the inspection itself become contested, thereby opening a second dispute about procedure? Did public statements arrive before internal documentation settled key facts? These questions do not solve the incident. They do reveal where the pathway is narrowing.

The quiet motif here is the file. It begins small: a report, a signal, an inspection note, a seizure record, a legal complaint. Then other documents gather around it - diplomatic messages, press statements, parliamentary questions, industry protests, legal briefs. The file thickens faster than the original evidence. That growth is the process one must learn to see. A fisheries crisis often begins not when force is used at its highest level, but when the file becomes too crowded with meaning for a narrow technical resolution to survive.

The broader lesson is that incident pathways are part of governance capacity. A state that recognises them early can choose language, timing, documentation, and procedural routing more carefully. A state that ignores them may still be legally correct and strategically clumsy. The same is true for fleets and coastal actors. Once a detention, seizure, or inspection is interpreted as a national affront, the possibility of treating it as an administrative matter begins to shrink. That is why the crucial analytic move is not to ask whether an incident is serious in the abstract. It is to ask where along the pathway it sits, which new audience has just entered, and what the next likely escalation point will be if nothing diverts it.

The unresolved question, then, is not whether the next dispute will begin with some small and familiar event. It almost certainly will. The harder question is which moment should count as the real warning sign: the first alleged crossing, the first contested inspection, the first visible detention, or the first point at which the domestic audience begins to hear not a fisheries case, but a story about humiliation, resolve, and the need to answer back.

Diplomacy, hotlines, and procedural off-ramps

In January 2019, the dispute over the Russian trawler *Melkart-5* in waters around Svalbard showed how quickly a fisheries incident can gather political meaning once inspection, escort, and protest begin to feed one another. Yet what kept that episode from becoming something larger was not agreement on principle. It was the continued existence of channels through which officials could still talk, compare accounts, and manage pace. The call matters.

That is the quiet importance of diplomacy in fisheries conflicts. It does not always resolve the underlying quarrel. Often it does something more modest and more urgent. It prevents a disputed boarding, detention, or boundary allegation from becoming the only language available to the parties. A hotline, a working-level contact, a technical meeting, or an agreed timetable for document exchange does not erase disagreement. It moves disagreement into a form that buys time, lowers emotional temperature, and narrows the chance that the next move at sea will also need to serve as the next political message. In contested fisheries, that is not procedural clutter. It is a strategic asset.

The basic mechanism is simple. A vessel is stopped, warned, or accused. One side believes the event proves unlawful conduct; the other says the authority exceeded itself or misunderstood the facts. Without an established channel, each side has strong incentives to speak first in public and interpret later in private. With an established channel, the immediate argument can be routed into notes verbales, inspection records, track comparisons, and timetables for further review. The name for this is procedural displacement. The explanation is that officials can often contain a dispute by changing its medium from confrontation to documentation. The implication is practical. A case that enters paper early is less likely to require theatre at sea to keep it alive.

This is why hotlines matter more than their plain name suggests. A hotline is not just a telephone number or emergency contact. It is a standing promise that an incident can be reported, clarified, contested, and managed before delay or rumour begins to do the work of escalation. In the context of distant fisheries, where conditions are harsh and reactions are often shaped by uncertainty, that promise matters enormously. A patrol commander who can speak to a counterpart directly may avoid treating a contested manoeuvre as defiance. A ministry informed quickly may choose caution in public because it knows that factual clarification is already under way in private. The show is quiet: a call placed, a message relayed, a timeline agreed. The implication is large. An incident that might otherwise harden into symbol remains, for a while longer, a problem to be processed.

The Joint Norwegian-Russian Fisheries Commission and its wider pattern of routine contact offer a useful example here. Its value has never rested solely on final agreement over every matter. The deeper achievement has been to create habits of communication strong enough that tensions over shared stocks can often be handled within an existing procedural frame rather than through improvisation under pressure. This does not make conflict disappear. It means conflict arrives in a room already furnished with working relationships, technical vocabulary, and known routes for clarification. That is the essence of a procedural off-ramp. The vehicle may still be moving too fast. At least the road exists.

A second example lies in the 2018 Central Arctic Ocean agreement, whose significance was not only precautionary restraint but the creation of a framework for consultation and joint scientific work before a full commercial fishery had taken shape. That arrangement recognised a fact many fisheries disputes confirm too late: if channels are built only after a confrontation begins, they are more likely to look tactical and insincere. If they exist beforehand, even minimal contact can carry more stabilising force because neither side has to invent procedure in the middle of a politically charged season. A disputed signal, sighting, or report can be fed into a known process rather than into immediate public accusation.

This is also why technical meetings matter, even when outsiders dismiss them as bureaucratic delay. A technical meeting can do several kinds of work at once. It can depersonalise a dispute by moving it from crews and ministers to specialists. It can narrow the question from grand principle to particular facts: timing, location, gear use, reporting sequence, interpretation of a measure. It can create a formal pause in which both sides refrain from harder steps while evidence is exchanged. Above all, it can allow governments to tell domestic audiences that they are not conceding anything, merely insisting on procedure. That is often enough to keep a fishing season governable. Procedure buys time. Sometimes time is the substance.

Interim measures serve the same purpose in a more visible form. Where a stock is moving, a season is short, or a detention has already sharpened attention, states may agree to temporary catch limits, reporting requirements, restricted access zones, or reciprocal restraint while the larger disagreement remains unresolved. Such arrangements are not elegant, and they often frustrate those who want clarity. Yet they can be deeply stabilising because they reduce the incentive to test the edge while preserving each side's formal position. The important point is that substantive agreement is not always necessary for practical de-escalation. A fishery can be managed provisionally long enough to prevent a process of escalation from becoming self-sustaining.

A reasonable limitation must be admitted. Proceduralism is not a cure-all. It can become a way of deferring difficult political choices indefinitely, especially if one side benefits from delay or if the underlying distributional conflict is too sharp to absorb through meetings and paperwork alone. A hotline that is never used honestly, a technical group that merely rehearses national talking points, or an interim measure that no one expects to endure can all become part of the problem rather than part of the solution. This is the danger of empty process. It looks calming from a distance while storing resentment underneath.

That limitation matters because de-escalation works only when procedure retains credibility. The parties do not need to trust one another warmly. They do need to believe that documents will be exchanged, timelines broadly kept, and channels used for clarification rather than performance. Once that belief weakens, even good mechanisms begin to fail. A delayed response is read as strategic. A technical meeting is treated as theatre. A proposal for interim restraint is heard as a trap for precedent. The same channels that once cooled the fishery then lose authority precisely because broader political relations or domestic pressures have colonised them. The call matters again, this time because its tone reveals what the institution can no longer absorb.

This is why de-escalation should be treated as something designable rather than as a matter of goodwill. Well-designed off-ramps usually share certain features: clear points of contact, agreed formats for exchanging evidence, understood timelines, distinctions between operational and political communication, and enough institutional memory that the first hours of an incident are not spent inventing a process from nothing. These are not glamorous features. They are the practical conditions under which a boarding, warning, or contested inspection can be prevented from becoming a national drama by default. Design matters because the sea offers little forgiveness for delay improvised under stress.

The diplomatic value of procedure also lies in what it allows states not to say immediately. Public ambiguity can be dangerous, but it can also preserve room for correction when facts are still uncertain and domestic audiences are listening for signs of weakness. A government with a trusted private channel can afford more restraint in public because it knows the matter is being handled somewhere credible. A government without that channel is more likely to perform resolve aloud because silence feels too risky. In this sense, procedural off-ramps do not merely manage disputes after they begin. They shape the kind of public language states can survive using while the file is still incomplete.

The broader lesson is that fisheries diplomacy often succeeds not by producing agreement, but by delaying the moment when disagreement must be expressed through forceful acts at sea. That sounds modest until one considers the alternatives. A fishery conducted at the height of seasonal pressure, under weak communication, and with domestic audiences already inflamed does not need much to tip into crisis. A routine call unanswered, a document withheld, a technical meeting delayed, an interim measure rejected too publicly: these are the points at which procedure stops acting as a cushion and starts being replaced by signalling. Then every next move grows heavier.

The unresolved question, then, is not whether hotlines, meetings, and interim arrangements can sometimes cool a dispute. They plainly can. The harder question is when procedure is still functioning as a real off-ramp and when it has become only a slower road to the same confrontation - a test visible in the first missed call, the first delayed document, or the first moment when each side begins to use the channel less to clarify facts than to prepare blame for the season that is already slipping away.

Strategic ambiguity and signalling

In January 2019, the dispute over the Russian trawler *Melkart-5* in waters around Svalbard showed how a fisheries incident can become politically charged without any side formally declaring a larger confrontation. The vessel was inspected. The handling of the case became contested. Officials spoke carefully. The legal and operational arguments remained partly open even as each move was watched closely for what it implied about jurisdiction, restraint, and resolve. Nothing in this sequence required martial language to feel serious. A pause can speak loudly.

That is why strategic ambiguity matters so much in fisheries governance. States do not always clarify everything they believe, everything they are willing to do, or everything they think a particular incident means. Sometimes they choose not to because clarity can narrow room for negotiation, force a domestic political line before the facts are fully checked, or create precedents that later prove awkward in adjacent disputes. Ambiguity, in this setting, is not simple confusion. It is often managed uncertainty. The question is not whether states dislike clarity in principle. It is whether, at a given moment, precision would cost more than it saves.

This does not mean ambiguity is harmless. The same uncertainty that preserves diplomatic space can also invite testing. The show is familiar enough: a vessel is boarded or warned, but the public language around the event remains measured and incomplete; an escort occurs, yet officials stop short of maximal rhetoric; a disputed crossing is neither fully conceded nor fully escalated; a ministry delays public comment while technical channels remain busy in private. The name for this is strategic ambiguity. The explanation is that governments are trying to hold together several goods at once - legal position, negotiating flexibility, domestic reassurance, and freedom from premature commitment. The implication is double-edged. Ambiguity can cool a dispute by preventing immediate hardening. It can also encourage brinkmanship by leaving room for operators and rival states to believe that caution is weakness.

The fisheries context sharpens this tension because enforcement choices are unusually legible. A boarding is a visible act. So is the decision not to board. An escort suggests one level of seriousness; a warning another; a quiet request for documentation another again. Publicising an inspection can satisfy domestic audiences and signal resolve. Keeping it low-key can preserve room for later technical settlement. Each choice communicates, whether intended or not. This is why communication in fisheries disputes should not be treated as commentary added after real action. The communication is part of the action. The manner, timing, and framing of enforcement decisions shape deterrence just as surely as the legal basis for the decisions themselves.

The 2019 Svalbard dispute remains instructive because it revealed the dense overlap between legal interpretation and signalling. A fisheries enforcement act was not read only as a fisheries enforcement act. It was read through larger questions about authority in the area, tolerance for non-compliance, and willingness to press matters beyond routine administration. Norwegian restraint in language, Russian protest, and the wider management of the incident all illustrated a truth often missed in casual debate: states may choose carefully calibrated public ambiguity not because they are unsure what happened, but because saying too much too early would convert one case into a precedent or a test of prestige they would rather postpone. The file stayed open.

A second example lies in the long history of the Cod Wars between Iceland and the United Kingdom. Those disputes were far more overt and politically charged than many contemporary fisheries incidents, and their legal and geopolitical context was distinct. Still, they offer a useful lesson about signalling without militarising the analysis. Patrol patterns, escort practices, language choices, and the degree of publicity attached to confrontations all communicated positions beyond the narrow act at sea. Ambiguity was sometimes preserved to keep negotiation possible, sometimes narrowed to satisfy domestic expectation, and sometimes exploited by those seeking to force a sharper line. What mattered was not merely the physical encounter but how that encounter was framed as either routine defence of fisheries rights or a larger affront requiring answer. The same act could be made to travel under different names.

This is why states sometimes tolerate ambiguity deliberately. The first reason is precedent. A government may believe that a fully explicit legal characterisation of one incident will bind or embarrass it later in another setting where the facts are similar but the interests differ. The second is negotiating space. A ministry that speaks too crisply in public can make compromise look like retreat. The third is domestic management. Ambiguity allows governments to reassure multiple audiences at once: officials can tell one group that the matter is under control, another that rights have not been surrendered, and another that technical review is still under way. These are not signs of unseriousness. They are ordinary techniques of political survival in disputes where the fishery, the law, and the national story are all moving at different speeds.

Yet the costs of such management are real. Risk-tolerant operators learn quickly from silence and from partial response. If a state repeatedly avoids strong public framing, delays decisive interpretation, or chooses not to attach legal consequences clearly to certain kinds of boundary testing, vessels and fleets may infer that the margin for assertive behaviour is wider than it really is. This does not require anyone to announce a policy of leniency. The lesson is drawn from pattern. A boarding not publicised, an escort not followed by visible sanction, a protest note not backed by operational change - these can all be read as signals that the state prizes flexibility more than immediate deterrence. Sometimes that reading will be wrong. It may still shape behaviour.

This is where signalling becomes inseparable from enforcement design. A state that boards early and publicly is saying something about its tolerance for uncertainty. A state that prefers documentation requests and port scrutiny is saying something else. A state that escorts but avoids seizure, or seizes but avoids public spectacle, is not acting neutrally. It is choosing what kind of message to send to fleets, rivals, domestic constituencies, and sometimes to future versions of itself. The important point is not to inflate these choices into military doctrine. That would distort the fisheries setting. The point is to recognise that communication and deterrence interact continuously in ordinary regulatory practice. One does not need gunboats and ultimatums to have signalling. A press release will do.

A useful limitation should be admitted. Not every ambiguous statement is strategic, and not every carefully calibrated operational choice is a deliberate signal. Governments are often genuinely uncertain in the first hours of an incident. Facts may be incomplete. Different agencies may disagree. Safety concerns may force operational caution that outsiders misread as political equivocation. This matters because over-reading every silence or partial statement can be as misleading as ignoring the communicative dimension altogether. The challenge is interpretive discipline. One must ask whether ambiguity is persistent, patterned, and institutionally useful to the actor deploying it, rather than simply assuming cunning in every pause.

That caution, however, does not weaken the larger argument. In fisheries disputes, ambiguity becomes strategic when it helps manage incompatible imperatives that cannot yet be reconciled openly. It lets governments avoid foreclosing negotiation while still marking concern. It allows technical channels to work before public language hardens. It keeps options alive during peak fishing seasons when immediate clarity may be impossible or too costly. At its best, this is stabilising. It prevents overcommitment and reduces the chance that routine enforcement will be read as an irreversible challenge. At its worst, it creates a permissive fog in which brinkmanship flourishes because no one quite knows whether the next step will be tolerated, protested, or answered.

This is one reason why signalling through publicity matters so much. To publicise an incident is to invite domestic audiences into the dispute before procedural channels have done their work. To suppress or soften publicity is to preserve administrative space, but also to risk looking evasive or weak if the story emerges from another source. States must therefore choose not only how to act, but how to narrate action. A low-key boarding, a quiet escort, an unadvertised inspection report, or a statement promising review rather than pronouncing judgment can all serve de-escalatory purposes. But they also shape the expectations of crews, industry organisations, and rival officials. The public frame is not a mirror. It is part of the operating environment.

The quiet motif here is the pause. A pause before boarding. A pause before comment. A pause between inspection and publication. A pause can be prudent. It can also be productive in the darker sense, giving risk-tolerant actors room to push again. In fisheries politics, pauses are rarely empty. They are read closely for tone, sequence, and what they suggest about future tolerance. This is why ambiguity cannot be understood merely as the absence of clarity. It is a managed interval in which meaning remains contested and therefore politically usable.

The broader lesson is that states communicate through method as much as through words. A fisheries authority that relies on remote verification and port checks communicates a different understanding of acceptable risk from one that privileges at-sea intervention. A ministry that releases detailed factual statements early communicates differently from one that insists on private review before public language. Neither path is inherently superior. Each organises deterrence and escalation differently. The serious analytical task is to see how communication choices shape the field in which fleets and rival states decide what they can test without forcing a response too costly to invite.

The unresolved question, then, is not whether ambiguity has uses. It plainly does. The harder question is where the line lies between ambiguity that preserves negotiating room and ambiguity that teaches operators and rivals that the state is reluctant to turn concern into consequence - a line often revealed only when the next boarding, escort, or carefully delayed statement is read not as prudence, but as permission.

Media, nationalism, and narrative hardening

In August 2013, when the European Union imposed measures against the Faroe Islands during the herring and mackerel dispute, the political struggle did not unfold only in negotiating rooms or legal texts. It unfolded in headlines, televised interviews, industry statements, and the familiar compression by which a complicated quarrel over stock distribution, procedure, and bargaining became a simpler drama about rights denied and fish taken. The details did not disappear. They were outrun. A phrase such as "our fish" or "stolen fish" could travel farther in a day than a technical clarification could travel in a month.

That is the proper starting point for thinking about information environments in fisheries governance. Disputes at sea do not move directly from vessel to ministry and then to settlement. They move through a public arena that rewards imagery, conflict, and moral clarity more readily than procedural nuance. A detained crew, a seized net, a photograph of officers on deck, or an accusation of intrusion can become the decisive fact in political life even while the underlying evidence remains contested. The picture arrives first. The file follows later.

This is why media and narrative should be treated as part of enforcement capacity rather than as something external to it. The show is familiar enough: an incident occurs, one side releases a brief statement or image, journalists compress the event into a recognisable frame, fishing communities supply emotional and economic context, opposition politicians search for weakness, and domestic audiences begin to judge the dispute before the relevant agencies have finished comparing tracks, reports, and legal authorities. The name for this is narrative hardening. The explanation is that public framing narrows the range of later compromise by turning a fluid incident into a fixed moral story. The implication is severe. Once the story hardens, even a technically sound off-ramp may begin to look like surrender.

The fisheries context makes this especially potent because the symbolic material is already close at hand. Crews are easy to personalise. Harbours are easy to photograph. Fish are easy to nationalise. A boundary allegation may be legally complex and politically manageable while it remains on a chart. It becomes something else once a family member appears on local television or a fisheries organisation says a crew has been humiliated or a national resource stolen. This does not make such reactions irrational. It means that they belong to the ordinary politics of disputed livelihoods. A state that ignores this will often discover that it has handled the law competently and the conflict badly.

The 2013 Faroese dispute is useful precisely because it showed how quickly external enforcement pressure could be reframed domestically as a larger question of recognition, autonomy, and fairness. Trade and port measures may have been defensible within one institutional logic, but their public meaning in a fisheries-dependent polity was never likely to remain technical. The dispute became a test of whether a small community's claims about changing stock distribution would be acknowledged or disciplined. Once the argument took that shape, compromise grew harder because any concession risked being read not as prudent adjustment but as public yielding under pressure. Narrative did not decorate the dispute. It reorganised it.

A second example lies in the recurrent imagery surrounding detentions or contested boardings in northern waters, including the 2019 dispute over the Russian trawler *Melkart-5* around Svalbard. Whatever one made of the underlying legal and operational issues, the political force of the incident depended in large part on what could be shown and said before the technical record had fully matured. A contested inspection is one thing in a report. It is another when it becomes a sign of selective enforcement, unwarranted defiance, or strategic provocation in public discourse. What matters here is not merely whether a government or fleet is right. It is whether the first available narrative leaves room for later complexity.

That room is often narrower than officials imagine. Ministries tend to think that facts, once assembled, will gradually discipline exaggerated claims. Sometimes they do. More often the first framing exerts disproportionate influence because it answers a simpler question than the file can answer quickly: who was wronged? A detained crew looks like wronged people. A gear seizure looks like injured livelihood. A phrase such as "stolen fish" or "our waters" provides immediate orientation in a way that zonal attachment, mixed-stock uncertainty, or evidentiary review never can. Narrative hardening therefore begins not when facts are falsified outright, but when one interpretive frame becomes emotionally and politically convenient enough that later corrections sound evasive rather than illuminating.

This is why nationalism enters fisheries disputes so readily. It need not arrive with overt jingoism. It can emerge more quietly through the pairing of resource, place, and dignity. A fish stock becomes more than a biological population; it becomes proof that a nation, region, or island community is being treated fairly or not. Once that transformation has occurred, media framing acquires a second layer of power. It no longer reports on a regulatory incident alone. It reports on whether the state is defending its people, whether outsiders are overreaching, whether local labour is being disrespected, whether national rights are being bartered away by timid officials. These are hard questions to answer with caution while a season is underway and the cameras are already pointed at the quay.

A useful limitation should be admitted. Media attention is not always destructive, and public scrutiny can serve valuable governance purposes. Authorities that would prefer to bury weak procedure, selective enforcement, or politically convenient ambiguity may behave more carefully when they know journalists and local communities are watching. Public pressure can force agencies to explain themselves, reveal inconsistencies, or treat crews more seriously. It can also widen the circle of accountability beyond the officials and industry actors most directly involved. The problem, then, is not publicity in the abstract. It is the kind of publicity that hardens before verification and that rewards maximalist framing over provisional honesty.

That distinction matters because governments often respond badly to narrative pressure. One common error is silence. Officials withhold comment while evidence is being reviewed, hoping caution will preserve flexibility. In the meantime, other actors supply the meaning. A second error is defensive over-clarity: a government, pressed by domestic criticism, declares a hard line before the case is mature, thereby reducing its own room for later adjustment. A third is technocratic abstraction: officials offer correct but bloodless explanations that fail to address the human and symbolic dimensions already driving public reaction. None of these responses is trivial. Information environments punish hesitancy, but they also punish clumsy certainty. Early communication is therefore not an optional public-relations accessory. It is part of the enforcement toolkit.

What, then, counts as credible communication practice in a fisheries dispute? Not propaganda, and not theatrical certainty. What works best is often narrower and more disciplined: early factual acknowledgment that an incident occurred; clear statement of what is known and not yet known; visible commitment to procedure; avoidance of language that assigns final blame before technical review; and enough specificity that domestic audiences do not feel they are being managed with platitudes. The point is not to win the argument in an hour. The point is to stop the argument from hardening around the most inflammatory interpretation available. A prompt, modest, verifiable account can preserve more room for diplomacy than either silence or chest-beating.

This is especially important because media dynamics interact with the procedural off-ramps discussed earlier. A hotline, technical meeting, or interim arrangement can cool a dispute only if public framing has not already rendered such measures politically toxic. Once the dominant story says the nation has been humiliated, a joint fact-finding effort may sound like capitulation. Once a detained crew is widely presented as victims of arbitrary authority, a routine administrative timetable may look cold or evasive. Information environments are thus not background weather. They condition whether de-escalatory design can still operate. A government may build sensible channels and still lose control of the dispute if it lets the public story outrun those channels beyond recovery.

The quiet motif here is the image. A vessel under escort. A crew waiting on deck. A stack of nets. A map with a line drawn thickly enough to look obvious. The image simplifies. That is its strength and its danger. Fisheries governance, by contrast, often depends on preserving just enough complexity that a dispute can be slowed, documented, and managed without every act becoming a referendum on national honour. When images harden faster than documents, governance becomes thinner even if the law remains sound.

This is why early narrative management should be understood as a matter of statecraft rather than spin. It requires preparation before the incident, not merely improvisation after it. Agencies need agreed lines of communication, protocols for rapid factual release, an understanding of which imagery or language will inflame matters unnecessarily, and enough co-ordination across fisheries, coast guard, foreign affairs, and political offices that contradictory messages do not create new controversy on top of the original one. The best time to design these practices is before a vessel is detained, a boarding contested, or a fish stock turned into a national symbol. Once the symbol is in circulation, the options narrow.

The deeper lesson is that media and nationalism do not merely distort governance from outside. They are part of the environment within which governance must prove itself credible. A state that can patrol, inspect, and prosecute but cannot communicate early and convincingly may still find its practical options collapsing under public pressure. A state with thinner physical capacity but stronger narrative discipline may preserve more room for negotiation because it prevents the first account from becoming the only account. Capacity, in this sense, includes informational poise.

The unresolved question, then, is not whether public framing matters. It plainly does. The harder question is which moment should count as the real beginning of a fisheries crisis: the detention itself, the first image that escapes into public view, or the first failure by an authority to fill the information space with something credible enough that "stolen fish", humiliated crews, and wounded sovereignty stop sounding like competing claims and start sounding like settled fact.

Dispute settlement options and their limits

In 1993, the International Court of Justice delivered its judgment in the Jan Mayen case between Denmark and Norway, addressing maritime delimitation in waters where fisheries mattered enormously to both sides. The judgment clarified a legal question. It did not make fish stop moving, nor did it remove the need for continuing political management between the parties once the law had spoken. That is the right place to begin, because it captures a truth often obscured by the rhetoric of resolution. A court can settle an argument about entitlement more cleanly than it can settle the life that follows the ruling.

This is the tension at the heart of fisheries dispute settlement. States want finality, but fisheries often live in conditions that resist final forms. Stocks shift, seasons tighten, domestic pressures fluctuate, monitoring remains uneven, and the parties continue to meet one another on the water and in markets long after a tribunal or arbitral panel has finished its work. Formal processes therefore offer something both valuable and incomplete. They can clarify rules, narrow ambiguity, and impose structure on competing claims. Yet they can also intensify stakes, consume time that seasonal fisheries do not have, and force positions into harder shapes than practical co-operation can easily absorb. A judgment can close one door. The fishery still opens the next morning.

Arbitration attracts states because it promises a controlled route to legal clarification. The show is familiar enough: parties frame claims carefully, select or accept a forum, submit memorials, argue over law and fact, and invite a third party to pronounce on a defined issue. The name for the attraction is bounded authority. The explanation is that arbitration can offer a more tailored, less publicly theatrical path than broader litigation, while still producing an outcome with legal weight. The implication is substantial. Where states cannot agree on a line, a competence question, or the interpretation of rights, arbitration can prevent the quarrel from being fought indefinitely through unilateral acts and accumulating resentment.

And yet arbitration changes the quarrel even as it channels it. Once a dispute enters formal legal process, each side has reason to sharpen its position, preserve consistency, and avoid gestures that might later look like concession. What might have remained a negotiable grey area can become a matter of pleaded principle. A fisheries issue that was once partly distributive and partly practical is translated into legal categories that do not always fit the seasonal and adaptive character of the resource. This is not a flaw in arbitration. It is a consequence of its strength. Legal precision requires bracketing some of the messy pragmatism by which fisheries are often actually governed. The parties gain clarity. They may lose room.

Adjudication before standing courts has similar virtues and similar costs. A court such as the International Court of Justice or the International Tribunal for the Law of the Sea can supply legitimacy, procedural rigour, and a sense that the result rests on more than bargaining power. In disputes where rules are genuinely unclear or where a state believes its rights are being persistently ignored, that can be immensely valuable. A court can clarify what the law permits, what it forbids, and where later negotiation must begin from a cleaner legal baseline. The Jan Mayen judgment remains instructive for precisely this reason. It did not solve all the political work of managing northern fisheries, but it did help stabilise the legal frame within which that work would continue.

Still, courts are poorly suited to some of the qualities that make fisheries disputes difficult. Seasonal pressure does not wait politely for pleadings, hearings, and judgments. A migrating stock may alter the practical significance of a legal issue while the case is still making its way through procedure. A domestic audience watching an active fishing season may treat recourse to adjudication as either principled resolve or abdication of practical responsibility, depending on the political mood. And once a case is before a court, the winner may gain law and lose flexibility, while the loser may lose law and become more brittle politically. A legal victory does not always produce a governable fishery. Sometimes it produces a better brief and a worse relationship.

Negotiated settlement remains the most common route not because states are sentimental about compromise, but because they often need arrangements that formal law cannot easily supply in time. The show here is more fluid: bilateral talks, joint commission meetings, side understandings, interim quotas, linked concessions, technical adjustments, and carefully worded formulas that allow each side to claim continuity while accepting some change. The name for this is managed accommodation. The explanation is that fisheries disputes usually involve more than law: they involve markets, ports, patrols, domestic audiences, scientific uncertainty, and the fact that the parties must keep dealing with one another when the season returns. The implication is why negotiated settlements, however unsatisfying to those who crave legal neatness, remain indispensable. They are often the only form of settlement elastic enough to keep the fishery functioning while disagreement persists.

A second named example helps make this plain. The 2018 Agreement to Prevent Unregulated High Seas Fisheries in the Central Arctic Ocean was not a classic settlement of a mature access dispute, but it was very much a dispute-management instrument. The parties did not wait for full commercial conflict and then seek a tribunal. They built a regime of restraint, consultation, and joint scientific work before the fishery existed at scale. That choice reflected a practical judgement: in some settings, it is wiser to manage uncertainty adaptively than to seek definitive legal closure too early. The agreement's force lay in its interim logic. It treated not-yet-settled questions as something to be governed provisionally rather than dramatically resolved all at once.

This is why so many fisheries disputes are managed rather than solved. "Managed" can sound weak, as though the parties lacked seriousness or courage. Often it signals realism. A dispute over moving fish in contested or shifting waters may have no final settlement that remains politically and ecologically adequate for very long. A quota formula that works for a few years may strain under later stock movement. A legal clarification may answer one jurisdictional question and leave a distributive one untouched. A negotiated settlement may preserve peace while leaving deeper fairness grievances unresolved. Management is therefore not the absence of settlement. It is the recognition that some disputes require recurring adjustment because the object of dispute does not stay still and the parties cannot withdraw from one another's presence.

A useful limitation should be admitted. Not all management is wise, and not all interimism is stabilising. Sometimes states use provisional arrangements simply to postpone difficult choices, hoping that ambiguity will serve them longer than clarity would. Sometimes a "temporary" measure becomes a recurring source of frustration because it never quite fits current realities but remains politically easier than redesign. Sometimes formal dispute settlement is avoided not because it would be too rigid, but because one side benefits from keeping the law uncertain. Managed disputes can therefore become stagnant disputes. The fact that many fisheries conflicts are best handled adaptively does not mean every avoidance of adjudication is prudent.

This is where trade-offs become most visible. Arbitration and adjudication can clarify rules, but may take too long, harden positions, or deliver answers too blunt for seasonal management. Negotiated settlements can preserve flexibility, but may encourage repeated brinkmanship if actors believe enough pressure will always reopen the bargain. Interim measures can buy time, but time can also be used strategically by the party most comfortable with ambiguity. The serious question is never which tool is abstractly superior. It is which mix of tools fits the temporal rhythm of the fishery, the legal uncertainty at stake, the domestic politics of the parties, and the degree to which continuing co-operation remains possible after the formal dispute has supposedly ended.

The motif here is the knot. Formal dispute settlement often promises to cut it. Negotiated settlement more often loosens it. Interim governance sometimes prevents it tightening further. In fisheries, cutting the knot may be satisfying in law and disruptive in practice; loosening it may be politically survivable and conceptually untidy. States therefore spend much of their time deciding not only how to defend their rights, but how much knot they can live with while the fishery remains operational. This is not indecision. It is a form of constitutional judgement about what kind of order is possible among mobile stocks, fixed jurisdictions, and recurrent seasons.

Adaptive, interim governance matters because it respects the fishery's tempo. A seasonal stock does not wait for the ideal jurisprudential moment. Crews need guidance before the next opening. Regulators need a rule, however temporary, that can be enforced without inviting collapse. Scientists need arrangements that preserve data exchange and mutual access. Markets need enough predictability that a dispute does not become a standing uncertainty premium on all commercial decisions. Interim governance provides these goods even when substantive agreement is absent. It keeps the quarrel from becoming the only thing the fishery knows how to do.

The deeper lesson is that settlement should not be confused with peace, nor management with failure. A formal ruling can sharpen the next dispute if it leaves domestic humiliation or practical misfit in its wake. A provisional bargain can reduce immediate risk and still require constant maintenance to remain legitimate. Fisheries law and fisheries politics therefore meet most honestly in the recognition that many conflicts must be carried, not cured. The best available institutional design is often one that clarifies enough, adapts enough, and restrains enough, without pretending that the sea will honour the finality that lawyers and politicians understandably crave.

The unresolved question, then, is not whether arbitration, adjudication, or negotiation can each serve a purpose. They plainly can. The harder question is when a dispute needs a definitive answer and when it needs a governable pause - and whether the parties can tell the difference before the search for final settlement produces a legal success too slow for the season, too rigid for the stock, or too sharp for the relationship that must still hold when the next migration begins.

Chapter 11

Food Security and Coastal Communities

What does it mean for a fishery to be successful if the export figures look healthy while the nearest coastal households feel less secure, less heard, and less able to feed themselves from the waters that have long shaped their lives? That question unsettles the tidy language of management because it forces two different measures of success into the same sentence. A state may point to order, value, and sustainability. A community may point to distance, exclusion, and thinning control over its own food future. Both may be describing the same fishery.

This tension is easy to miss when fisheries are discussed chiefly through quotas, patrols, markets, and disputes between states. Those matters are real and indispensable. Yet the sea is also intimate. It enters kitchens, work rhythms, seasonal memory, and the practical dignity of knowing that a place can still draw nourishment from nearby waters. Once that connection weakens, a fishery can remain well governed on paper while becoming politically brittle in the communities expected to live with its consequences. The boat returns. Not all value lands.

That is why food security cannot be treated here as a soft addendum to the harder questions of allocation and enforcement. It belongs near the centre. Governance failures often strike local nutrition and cultural practice before they become visible in aggregate catch records or trade flows. Rights that look minor from the perspective of industrial management may be major from the perspective of daily life. A licensing rule, a quota concentration, a landing decision, or an access barrier may appear administratively rational and still alter who eats fish locally, who continues to fish, and who begins to feel that the resource is no longer politically theirs. Those shifts matter because legitimacy begins close to home.

The central tension lies here: fisheries governance often promises efficiency, sustainability, and orderly allocation, but communities judge it partly by whether it preserves meaningful access, recognises small-scale dependence, and leaves room for local adaptation. The conflict is not between sentiment and seriousness. It is between different kinds of seriousness. States need systems that can regulate scarcity, manage moving stocks, and survive external pressure. Coastal communities need those same systems to remain socially intelligible and materially bearable. When the second need is ignored, the first often weakens in time. Dispossession has a politics.

The stakes reach beyond equity in the moral sense. A community that feels excluded from the fishery's gains, priced out by leasing, or reduced to watching "its" fish leave under rules it did not shape is less likely to trust the institutions asking for restraint, reporting, or sacrifice later. That mistrust does not stay local. It enters national politics. It affects who is believed when governments defend quota bargains, market restrictions, or adaptive measures in international settings. Domestic distributional design can therefore travel outward. States speak abroad with voices shaped partly by what they have done at home.

This matters later as much as now because adaptation is not evenly available. Some communities can diversify, reorganise, and bargain effectively when fish move or regulations tighten. Others cannot. Infrastructure may be fixed, capital scarce, and alternative work limited. Under those conditions, "adaptation" can become a polite way of describing who is expected to absorb loss. Local knowledge and political voice are therefore not decorative additions to governance. They are resources that can widen the menu of possible adjustment and prevent difficult change from becoming a straightforward lesson in who counts and who does not. Silence is expensive.

The path ahead follows that pressure where it is felt most acutely. It begins with the distinction between subsistence, small-scale need, and the larger industrial logic that can easily drown them out in law and policy. It then moves into the internal politics of quota allocation, where concentration and leasing can alter the social meaning of a fishery long before any international dispute appears. From there the argument turns to adaptation itself, not as a slogan of resilience but as a contest over infrastructure, participation, and who gets to speak with authority when communities are asked to adjust to a sea and a market that do not wait.

A useful caution belongs here. No community is frozen in innocence, and no small-scale claim is automatically just because it is local. Fisheries are full of internal hierarchies, exclusions, and conflicts over who really benefits from the language of tradition or rights. Romanticism would be as unhelpful here as technocratic indifference. The serious task is harder: to ask how institutions can preserve food security, cultural continuity, and practical access without pretending that every local arrangement is fair or that every industrial claim is inherently corrosive. Governance must choose. It should at least choose honestly.

The harder question, then, is not whether food security matters to fisheries politics. It plainly does. The sharper question is what kind of system remains durable when the fish are still being caught, the law is still being enforced, and the markets are still moving, but the communities living closest to the resource no longer believe that the fishery feeds them in any meaningful political sense.

Subsistence, small-scale fishing, and rights

In 1999, the Supreme Court of Canada issued its decision in *R v Marshall*, and a legal argument about treaty rights quickly became something larger: a public struggle over who gets to fish, on what terms, and whether law can recognise small-scale and livelihood-based claims without treating them as quaint exceptions to the real business of the sea. The case arose in Atlantic Canada, not the Arctic, but its force travelled widely because it exposed a common error in fisheries policy. Governments often speak as though all access can be translated into quota, effort, and market value. Communities living closest to the water know otherwise. A fish can be counted two ways.

That distinction is where the politics of subsistence and small-scale fishing properly begins. Industrial management systems tend to measure success in aggregate terms: total catches within limits, export earnings protected, fleets regulated, markets supplied. Those are serious goods. Yet they do not settle what matters to a household, a village, or a coastal Indigenous community whose relationship to fish is nutritional, social, and cultural before it is commercial. A fish landed for family use, ceremonial exchange, or local sale enters a different moral economy from fish moving through distant processors and export chains. When governance fails to respect that difference, it usually injures the most local uses first while leaving broader catch statistics looking deceptively calm. The bucket empties early.

This is why subsistence and small-scale needs should not be folded casually into the same analytic box as industrial quotas. The show is plain enough: a state keeps aggregate harvest stable, export routes remain open, and the formal management system appears orderly, yet local households find access harder, seasonal customary practices narrower, and nearby fish less available to them in practice. The name for this problem is distributive invisibility. The explanation is that management systems built around total volume and formal allocation often fail to see who can still reach the fish, under what costs, with what equipment, and through what rights. The implication is more serious than a mere fairness complaint. A fishery can look sustainable in aggregate while becoming socially brittle at its edges.

This brittleness matters because food security is not exhausted by calories or commerce. In many northern and coastal regions, fish are woven into routine nutrition, reciprocal exchange, local authority, seasonal identity, and intergenerational teaching. A governance failure that restricts or destabilises those uses may therefore strike cultural practice and local diet before it registers as a macroeconomic problem. A household that loses practical access to fish does not care much that national export performance remains sound. A community that must watch nearby catch move into distant markets under rules it did not shape will not be reassured by official claims that the stock itself is still well managed. The law may hold. The life around it may thin.

A second example helps clarify this point. Alaska's subsistence fisheries regime, for all its complexity and contestation, reflects an enduring recognition that local dependence cannot simply be absorbed into commercial logic without doing violence to political reality. The legal and administrative details are intricate, and the regime has generated repeated controversy over preference, access, and identity. Yet the underlying lesson is plain. Once a state acknowledges that some fisheries matter differently to those who eat from them, teach through them, and organise local life around them, it cannot pretend that all access claims are interchangeable. Doing so would not produce neutrality. It would privilege those best placed to translate fish into formal market power.

This is where rights become crucial. Rights in fisheries are often discussed as if they were primarily instruments for regulating efficiency, investment, and predictability. That is only part of the story. Rights also decide whose relationship to the fishery is recognised as politically valid. A subsistence right, a community access right, or a preferential small-scale allocation is not simply a technical departure from market logic. It is a statement about what kind of dependence the state considers worthy of protection. The show is visible: a category reserved, a season tailored, a local fishery shielded, or a co-management arrangement built around knowledge that does not enter quota models neatly. The explanation is that law is being used to preserve a social connection to the resource, not merely to optimise extraction. The implication is that rights, in this setting, help determine whether governance remains intelligible to those asked to live under it.

And yet rights themselves do not solve everything. A formal recognition of subsistence or small-scale fishing can coexist with underfunding, procedural delay, weak local infrastructure, or administrative rules that make practical access increasingly difficult. A right on paper may not survive long steaming distances, expensive fuel, changing species distribution, or licensing requirements designed with industrial operators in mind. This is a useful limitation because it prevents a lazy legalism. Declaring a right is not the same as making that right usable. A state can congratulate itself on formal inclusion while local people continue to lose actual access through costs, geography, or cumulative bureaucratic friction.

That practical gap is one reason equity matters so much for legitimacy. "Equity" is an abstraction until one asks the obvious questions. Who can still fish without taking on commercial levels of debt? Who can still bring home fish for household use without crossing rules built for export fleets? Who can still pass on customary practices to younger generations without discovering that the legal system recognises the tradition ceremonially but not operationally? These are visible tests. When communities answer them negatively, they begin to read management not as order but as dispossession. The state may still be acting lawfully. It has started to lose consent.

Consent matters because compliance is never only a function of surveillance and penalty. People are more likely to tolerate conservation limits, seasonal closures, and reporting demands if they believe the regime sees them as legitimate participants rather than as residual claimants to be accommodated after the "real" fishery has been allocated elsewhere. Where small-scale and subsistence actors feel systematically subordinated, rule-following becomes thinner and more grudging. That does not always mean open resistance. More often it means political withdrawal, distrust of scientific advice, low confidence in consultations, and a growing willingness to interpret every new measure as another step in a longer story of exclusion. Compliance weakens quietly before it fails noisily.

This is why food security must be linked to institutional durability. A regime that secures industrial order but undermines local nutrition and cultural practice may function for a time, especially if export revenues remain high and enforcement strong. Over longer periods, however, it becomes vulnerable to backlash from the very communities whose co-operation it needs. Coastal representatives harden their positions. Governments become more defensive domestically and less flexible internationally. Small disputes about access or timing start to carry the weight of much larger grievances. What looked like a narrow problem of local allocation begins to reshape the political climate in which broader fisheries governance must operate. The bucket empties early, and the argument arrives late.

A reasonable counterpoint should be faced directly. Not every claim made in the name of small-scale or subsistence fishing is automatically equitable, and not every restriction on local access is unjust. Communities themselves contain inequalities, gatekeepers, and struggles over who counts as local, traditional, or entitled. Some small-scale fisheries overexploit vulnerable stocks. Some appeals to cultural practice can be deployed strategically in ways that deserve scrutiny. Recognising this does not weaken the case for differentiated rights and serious attention to food security. It strengthens it by insisting that equity must be designed, not merely assumed. A state that romanticises local claims will govern poorly. A state that ignores them will also govern poorly, and usually more destructively.

The stronger argument is therefore institutional rather than sentimental. Durable fisheries governance needs categories and procedures capable of distinguishing between uses that are economically comparable and uses that are politically incommensurable. A tonne of fish landed into a freezer trawler's supply chain is not socially equivalent to fish landed into a household economy or a small coastal exchange network, even when the species is the same. Treating them as equivalent simplifies administration and distorts reality. Treating them as wholly separate may create new inequities and conservation problems. The serious work lies in building systems that can acknowledge difference without surrendering discipline. That is harder than a universal quota formula. It is also closer to the truth of how fisheries are actually lived.

This is where small-scale fishing reveals something larger about the state. A government that can articulate industrial policy but cannot protect local nutritional and cultural access is showing its priorities with unusual clarity. It is also narrowing the moral basis on which later sacrifices can be asked of coastal citizens. If a fishery no longer feeds or recognises nearby communities in any meaningful way, why should those communities trust the same institutions when they ask for patience, restraint, or international compromise? This is not a rhetorical flourish. It is a recurring political fact in resource governance. Those who feel least seen are often least willing to underwrite the system's legitimacy when it comes under pressure.

The quiet motif here is the bucket. It is ordinary, local, and easily ignored by managers thinking in tonnes, quotas, and markets. Yet it marks the place where fisheries policy becomes daily life. A full bucket means more than food. It means continuity, skill, reciprocity, and a sense that the nearby sea still answers, however imperfectly, to the nearby community. When that bucket stays empty while export shipments continue moving, no amount of aggregate reassurance can fully mask what has been lost.

The deeper lesson is that equity is not a charitable supplement to serious fisheries governance. It is one of the conditions under which governance remains believable. Subsistence and small-scale rights remind us that institutional durability depends partly on whether local people can still recognise themselves in the rules that order the fishery. A system that secures stability for markets while weakening that recognition may appear strong until the next crisis demands public trust it no longer possesses.

The unresolved question, then, is not whether industrial quotas and export systems have a place. They plainly do. The harder question is what kind of fishery a state believes it is protecting if aggregate catches remain stable while the earliest and deepest losses are borne by those for whom fish are still food, practice, and local proof that the resource has not been entirely translated into somebody else's paperwork.

Distributional justice in quota systems

In 1984, Iceland introduced its individual transferable quota system, and a technical reform of fisheries management gradually became one of the most searching political arguments in modern coastal society. The reform promised order, efficiency, and discipline in a sector long exposed to biological and economic strain. It also changed the social geography of access. A quota is never only a number. It is a map of who may stay.

That is the place to begin because quota politics is often discussed as though the central question were merely whether fish stocks can be conserved while fleets remain profitable. Those questions matter deeply. Yet quota systems also distribute time, risk, credit, bargaining power, and the right to imagine a future in the fishery. When that distribution grows too narrow, communities may remain surrounded by productive waters and still come to feel that the resource has moved away from them politically. The fish are there. The access is not.

This is why distributional justice belongs near the centre of quota analysis rather than at its sentimental edge. The show is familiar enough: quotas concentrate through purchase or inheritance, leasing becomes a permanent condition rather than a temporary adjustment, entry costs rise, and coastal fishers or younger crew members find that participation now depends less on skill and effort than on access to capital or to someone else's paper rights. The name for this is internal dispossession. The explanation is straightforward. A quota system that makes rights transferable and scarce will usually reward those already best placed to hold, aggregate, and finance them. The implication is larger than a fairness complaint. Community resilience begins to depend not only on the health of the stock, but on who can still turn that stock into local work, food, and voice.

Iceland remains the clearest example because it shows both the managerial appeal and the social cost of concentration. The quota system brought real advantages in planning, fleet rationalisation, and the capacity to align harvesting more closely with biological and commercial constraints. That is not trivial. Yet over time it also sharpened public anger about ownership, leasing, and the separation of fishing rights from the places that had historically lived by them. In some communities, the most painful fact was not that fish disappeared. It was that rights attached to fish could circulate away from the harbour while the harbour itself was left with the afterlife of an industry no longer fully anchored there. A permit can travel. A town cannot.

The mechanism is not complicated. Transferability improves flexibility by allowing quota to move towards operators who can use it efficiently. Leasing deepens that flexibility by letting rights holders and active fishers adjust to changing conditions. The same arrangements, however, can produce a class of fishers who work without owning durable access and a class of owners who hold durable access without living the daily risk of the fishery. The distinction matters. The name for the resulting structure is layered access. The explanation is that one group owns the future while another rents the present. The implication is corrosive for community resilience. A coastal settlement may remain busy in the short term while losing its ability to decide what comes next.

This is where justice becomes a practical governance issue rather than a moral abstraction. A resilient community needs more than current landings. It needs continuity of participation, plausible entry routes for younger people, enough locally rooted control that the fishery still feels like a shared institutional world rather than a distant asset class, and enough retained value that shocks do not immediately become collapse. Quota concentration weakens these conditions even when biological management appears sound. The fishery becomes less adaptable socially at the same time as it becomes more orderly administratively. That exchange may be defensible in some cases. It is never neutral.

A second example comes from Alaska, where community development arrangements and ongoing struggles over access, leasing, and local participation reveal the same underlying truth in a different institutional form. Alaska has not followed Iceland's exact path, and its legal, Indigenous, and regional structures differ in important ways. Yet the recurring political question is recognisable: how can a state design allocation systems that preserve industrial viability without turning coastal communities into spectators of value generated nearby but controlled elsewhere? The fact that this question keeps returning, under different legal architectures, suggests that it is not an accident of one national system. It is a structural tension in quota governance itself.

Leasing deserves special scrutiny because it often appears economically modest while proving socially decisive. A leased quota can keep a vessel active for a season, preserve local landing flow, and provide flexibility where permanent transfer would be too blunt. But a fishery that runs heavily on leasing may slowly institutionalise dependence. Active fishers bear operational risk while surrendering a growing share of value to rights holders positioned above the daily uncertainty of weather, price, and catch. The result is not always visible in aggregate data. It appears in the arithmetic of households, in thinner margins, in hesitation about investing in gear or crews, and in the quiet realisation that work continues while control drifts further away. Leasing can therefore stabilise participation and hollow it out at the same time.

Access barriers for new entrants intensify the pattern. A young fisher or a small operator confronting high quota prices, entrenched ownership, and a market organised around existing rights will often find that the formal openness of the system is misleading. The fishery is technically accessible, but only through debt, dependence, or inheritance. This has political consequences. A regime that offers no believable route into ownership or secure access teaches coastal communities to regard fisheries governance not as a common system but as a settled hierarchy. Once that perception deepens, legitimacy thins. Conservation rules begin to look less like collective necessity and more like obligations imposed by a structure already captured by others.

A sensible limitation should be admitted. Concentration is not always a sign of corruption or design failure. Some degree of consolidation can improve safety, reduce wasteful competition, and support better investment in vessels, quality, and planning. A fishery fragmented too finely may struggle to remain profitable or biologically disciplined. Likewise, not every complaint about quota distribution reflects pure dispossession; some local arguments defend inefficient or exclusionary arrangements of their own. This matters because distributional justice cannot mean freezing every community structure in place or treating every transfer as betrayal. The real question is not whether concentration exists. It is how much concentration a system can absorb before it begins to damage the social conditions on which durable governance depends.

Those social conditions do not stay within domestic borders. Internal inequality often spills into international posture because governments facing discontent at home become more brittle abroad. A minister negotiating quota shares, access arrangements, or enforcement co-operation does not arrive as a blank strategist. The minister carries the knowledge that coastal regions already feel excluded, that leasing controversies are politically live, that local representatives are watching for signs of surrender, and that another perceived loss could strengthen the claim that the national system no longer protects the communities it claims to manage. External firmness is often fuelled by internal fragility. This does not make international disputes insincere. It does mean that some of their emotional charge originates at home.

This spillover takes several forms. A government may resist compromise more strongly because domestic critics would frame any concession as further proof that remote holders, foreign fleets, or distant ministries have all profited while local communities have lost. A state may dramatise a moving-stock dispute because it needs to show that external pressure, rather than internal allocation design, is the source of coastal distress. A country may also become less able to sell sensible international agreements domestically because the public no longer trusts the fairness of the national distribution that will follow any deal. Domestic quota design, in other words, shapes not only who fishes, but how a state speaks and bargains when fish cross lines on the chart.

That is why distributional justice should be read as part of external stability. A country that has built a quota regime widely seen as fair enough, however imperfectly, possesses more room to negotiate, adapt, and explain change. A country whose domestic system is read as concentrating rights, socialising risk, and narrowing entry often lacks that room. Every international adjustment then enters a field already primed for grievance. The fish move north. The political anger finds the nearest available target, whether or not that target is the original cause of the underlying problem. External hardening can therefore be a symptom of internal design choices long in the making.

The quiet motif here is the key. A quota right acts like one. It opens the season, the credit line, the investment plan, the future. When keys accumulate in fewer hands, others may still work in the building, but they no longer decide who enters or what remains inside. Coastal regions know this viscerally. They may still see boats on the water and fish at the quay, yet sense that the real control has shifted elsewhere. A system can remain productive while losing the local confidence that once made it politically durable.

This is why community resilience cannot be separated from allocation structure. Resilience is visible when a town can retain enough value, decision-making, and opportunity to absorb biological shifts or market shocks without immediate political despair. It is visible when younger fishers can imagine entry, when leasing does not consume too much of the working margin, when quota ownership remains connected enough to place that public arguments about the fishery still carry some local confidence. The absence of resilience is equally visible: chronic dependence, resentment towards absentee ownership, defensive politics, and a growing readiness to interpret any external dispute as one more episode in a longer domestic story of being left behind.

The deeper lesson is that quota systems do more than divide fish. They divide expectation. They tell citizens which relationships to the sea are meant to endure and which are expected to adapt, rent, or disappear. A government that treats this as merely a technical matter of efficiency will eventually discover that it has built a fishery better at conserving biomass than at conserving belonging. That imbalance is politically expensive. It reshapes compliance, corrodes trust, and narrows diplomatic room precisely when changing seas demand more flexibility, not less.

The unresolved question, then, is not whether quota systems can be biologically and economically effective. They can. The harder question is what kind of polity they create when rights become concentrated, leasing becomes normal, and access hardens into privilege - and whether a fishery can remain stable for long if the communities closest to it increasingly believe that they are being governed by a resource they no longer meaningfully hold.

Adaptation strategies and political voice

In 2004, the Alaska Native Claims Settlement Act was already decades old, yet the political questions it left alive around community control, participation, and access to local resources had hardly faded. Across northern coastal regions, the formal language of rights had never fully settled the practical matter of adaptation: who could still change with the fish, who had the infrastructure to do so, and who was expected simply to endure the consequences of decisions made elsewhere. The harbour could remain busy while the community grew quieter.

That is the right place to begin because adaptation is too often discussed as though it were a neutral technical response to changing ecological and market conditions. In reality, adaptation is distributed unevenly. Some communities can diversify, experiment, reorganise labour, and press their claims effectively. Others face cold stores built for the wrong species, plants tied to declining fisheries, vessels too small or too specialised for new opportunities, and capital costs that turn the language of resilience into a disguised command to absorb loss. A strategy is only real if someone can afford it.

This is why adaptation should be treated as a political and institutional question, not merely an economic one. The show is familiar enough: a stock shifts, a quota regime tightens, a market changes, or a local landing pattern weakens. One town responds by widening its portfolio, building new partnerships, or entering co-management forums with some leverage. Another finds that its infrastructure, debt, labour force, or legal status leaves little room to move. The name for this is unequal adaptability. The explanation is straightforward. Communities do not confront change with the same tools, and governance often overestimates flexibility because it mistakes formal possibility for practical capacity. The implication is serious. A policy can urge adaptation and still produce dispossession if the means to adapt are absent.

Diversification is usually the first and most celebrated response. It sounds sensible because it often is. A community may shift effort towards a different species, expand into more than one fishery, combine fishing with processing or tourism, or build local value through direct marketing and smaller-scale supply chains. These adjustments can reduce dependence on a single stock and create room to survive ecological or regulatory shocks.

But diversification is not magic. It depends on vessel design, harbour facilities, freezing and storage capacity, available permits, local skills, and access to finance. A trawler built for one fishery does not lightly become a vessel for another. A plant designed for volume does not easily transform into a flexible, high-value processor. To say "diversify" is often to say "rebuild", and rebuild is expensive.

This is where co-management enters as something more substantial than a fashionable administrative term. In its strongest form, co-management allows local users, Indigenous authorities, scientists, and state agencies to share responsibility for rules, observation, and adjustment. The value of this arrangement is not only democratic sentiment. It is epistemic and practical. Local knowledge can reveal timing, distribution, gear effects, and community constraints that distant regulators or national quota systems routinely miss. The show is concrete: meeting tables that include people who fish the area, seasonal rules shaped by lived observation, enforcement priorities informed by local practice, and feedback loops that run faster than a ministry memo. The explanation is that governance improves when those closest to the fishery help define what counts as workable. The implication is that participation is not a concession to local feeling. It is a way of governing better under uncertainty.

A useful example comes from Alaska, where co-management arrangements involving Indigenous communities and state or federal authorities have long shown both the promise and the limits of local participation. The institutional designs vary, and many remain contentious, but the broad lesson is clear. Where local users have meaningful standing in management, the resulting rules are often more attentive to timing, access, subsistence needs, and practical enforcement than systems designed wholly from a distance. This does not guarantee harmony. It does, however, reduce the likelihood that adaptation will be presented as something done to communities rather than with them. That distinction matters enormously when ecological change begins to bite.

A second example can be found in Icelandic coastal politics, where local advocacy has repeatedly emerged in response to quota concentration, plant closures, and the social consequences of centralised fisheries design. Iceland is often discussed through the lens of efficiency and ownership, rightly enough, but its politics also show how communities adapt not only economically but institutionally. They organise, litigate, lobby, build municipal alliances, and force questions of place and fairness into national debate. Advocacy here is not a decorative democratic ritual. It is a survival strategy for places whose economic future is shaped by rules they did not fully write. When capital is fixed and fish move, voice becomes part of the infrastructure.

That point deserves emphasis because political voice is often underestimated in fisheries analysis. Voice does not land fish or freeze fillets, but it shapes whether communities can bargain for the conditions under which adaptation becomes possible. A town that can press for landing obligations, infrastructure funding, adjusted access rules, community quota arrangements, or more flexible licensing has more room to respond than one left to absorb the decision silently. The abstraction here is participation. Its visible forms are council pressure, legal challenge, organised industry representation, Indigenous governance, public campaigning, and the ability to make ministers listen before a season is lost. The implication is plain. A community without voice is often a community asked to adapt on terms chosen elsewhere.

A sensible limitation must be admitted. Not all adaptation is desirable, and not every local demand is automatically wise or equitable. Communities can defend exclusionary arrangements, unsustainable practices, or politically protected claims that deserve scrutiny. Co-management can become tokenistic or captured by local elites. Advocacy can harden into nostalgia for a fishery that cannot return in the same form. This matters because the argument here is not that local preference should simply override ecological or national considerations. It is that local knowledge and political voice are governance resources, and systems that ignore them usually govern less intelligently and provoke more conflict than systems that try, however imperfectly, to incorporate them.

The most important limit, however, lies in fixed capital. Adaptation is hardest where vessels, docks, factories, roads, skills, and household economies were built around one fishery or one pattern of access. A plant cannot easily move inland or transform overnight. A family tied to debt, housing, and seasonal labour in one harbour cannot lightly follow a stock north or sideways into a different regulatory category. This is why community adaptation should never be romanticised as infinite improvisation. Some places can pivot. Some cannot. A government that treats them alike will misread both resilience and risk. What looks like stubbornness from a capital city may in fact be the realistic recognition that the available options are thin.

This is where conflict prevention enters the picture. Communities that feel heard, represented, and involved in adaptation design are more likely to accept difficult measures, temporary losses, and ecological uncertainty without reading every change as another episode of imposed decline. Communities that feel ignored are more likely to interpret the same measures through the language of abandonment, extraction, and disrespect. That shift in perception matters beyond the locality itself. It shapes national politics, hardens external negotiating positions, and reduces institutional trust across the fisheries system. Participation, in this sense, is not only democratic good practice. It is a preventative measure against the escalation of distributive grievance into broader political conflict.

The quiet motif here is the quay. It is where adaptation becomes visible. New species land there, or do not. Different gear appears, or does not. Trucks keep coming, or slow. Meetings in distant offices may define the policy, but the quay records whether the community could actually turn policy into a changed livelihood. A busy quay does not prove justice. An empty one does not prove mismanagement. But the quay tells the truth about whether adaptation remained a slogan or became a material possibility.

This is why local knowledge matters so much. It is not sacred because it is local. It matters because it can identify practical thresholds long before centralised systems notice them. Fishers know when a species is appearing differently, when gear assumptions no longer fit, when a closure makes sense biologically but not operationally, when a plant cannot handle a regulatory shift, and when a formal adaptation pathway exists only on paper. This knowledge is not sufficient by itself, but nor is scientific abstraction sufficient without it. Governance becomes more durable when it allows these forms of understanding to challenge, refine, and sometimes correct one another rather than forcing one into ceremonial deference to the other.

The deeper lesson is that adaptation and voice belong together. A state cannot plausibly ask communities to adjust to moving stocks, tighter allocations, and changing markets while denying them meaningful participation in the design of those adjustments. Where adaptation is possible, voice helps make it legitimate. Where adaptation is constrained, voice helps reveal the constraint before policy hardens around a fiction of flexibility. Either way, participation improves the state's odds of governing change without manufacturing unnecessary conflict around it.

The unresolved question, then, is not whether communities can adapt. Many do, impressively. The harder question is which communities are being asked to adapt with real tools, which are being asked to adapt with words, and whether a fisheries regime can remain politically stable for long if the places bearing the sharpest costs have the least power to define what adaptation should actually mean.

Chapter 12

Adaptive Governance for Moving Stocks

How does a state defend its rights without freezing itself into rules that no longer fit the fish? That is the governing paradox once stock movement becomes normal rather than exceptional. Fixed allocations offer certainty, but certainty can turn brittle when distribution changes faster than diplomacy. Flexibility sounds wiser, yet without trusted limits, it can look like drift, weakness, or disguised surrender. The question is not whether adaptation is desirable in the abstract. The question is what kind of adaptation can preserve order while the ecological ground keeps shifting under it.

This tension has haunted fisheries governance for years, though it is often described too narrowly as a technical matter of updated science or more frequent negotiation. The deeper issue is political design. A regime built for stable distributions can become an engine of grievance when fish move and the rules do not. A regime that adjusts too easily may calm one dispute only to create another by making every season feel provisional. The ruler on the desk matters here. It can measure change, but it can also become the object of argument once the parties stop trusting what counts as enough movement to justify a new bargain.

That is why adaptive governance should be treated as a security problem as much as an administrative one. When institutions cannot absorb ecological change, the pressure does not remain ecological for long. It passes into quota politics, enforcement strain, public rhetoric, and the tactical behaviour of fleets that learn quickly where ambiguity is profitable. A rigid system can provoke repeated confrontations because every shift must be fought as if it were unprecedented. A loose system can invite opportunism because no one is quite sure which line still holds. The hard task is to design enough movement into the rules that adaptation becomes ordinary rather than exceptional, without turning ordinary change into a permanent dispute over authority.

The stakes are durable and practical. Adaptive systems decide whether governments spend their time revisiting the same argument season after season or whether some of that conflict is absorbed before it reaches patrol vessels, courtrooms, and ministers' statements. They also decide whether agencies are surprised by new entrants, enforcement surges, or collapsing co-operation, or whether those developments arrive in a system already prepared to sort responsibility, information, and decision rights. Preparedness is rarely glamorous. It is often the difference between a tense season that remains governable and a tense season that teaches every participant to expect improvisation, delay, and blame.

This matters later as much as now because the political economy of compliance depends on whether the rules make lawful behaviour rational under changing conditions. If access fees, benefit sharing, verification rules, and adjustment procedures all remain tied to an older fishery, then grey tactics begin to look less like deviance and more like a rough form of adaptation. The problem is not moral failure. It is design failure. A regime that ignores behavioural response will eventually discover that enforcement is trying to compensate for incentives it should have corrected earlier, and that technology is being asked to settle disputes that norms should have prevented from hardening in the first place.

A further difficulty follows from trust. Adaptive governance sounds modest until one asks who measures change, who certifies a threshold has been crossed, who decides an interim response, and who verifies that the tools used to observe, record, and audit behaviour have not themselves become new arenas of contest. Instruments do not calm a fishery merely by existing. They calm it when the parties share enough confidence in how they are used that an adjustment can be treated as a rule working rather than as a manoeuvre disguised as procedure. That is why verification norms matter so much. The sensor, the database, and the audit trail are only as stabilising as the common practice around them.

The path ahead follows that institutional problem where it becomes most tangible. It begins with trigger rules and revisable allocations, where flexibility must be formal enough to reduce renegotiation fatigue without unsettling sovereignty claims. It then turns to scenario planning and institutional readiness, where the point is not prophecy but disciplined preparation for pressure. From there the argument moves into incentive design, where compliance becomes sturdier when the payoffs of lawful conduct are made more attractive than the returns from evasive tactics, before ending with the governance of technology itself, where shared standards for verification decide whether new tools narrow disputes or merely modernise them.

A useful caution belongs here. Adaptive governance is not a synonym for endless revision, nor is preparedness a cure for political mistrust. Some disputes are not softened by more flexibility because the real conflict lies in distribution, identity, or domestic constraint rather than in the absence of a clever rule. Some trigger mechanisms will fail because the measurements they rely on are contested. Some scenarios will never occur, and some that do occur will arrive in a form no planning exercise quite anticipated. None of this weakens the adaptive case. It clarifies that adaptation must be disciplined, not romanticised.

The harder question, then, is not whether moving stocks require moving institutions. They plainly do. The sharper question is what kind of institutional movement counts as strength rather than concession: a rule that bends before pressure, a plan that quietly assigns authority before the next surge arrives, an incentive structure that leaves fewer profitable shadows, or a verification culture sturdy enough that when the fish move again, the argument does not have to start from zero.

Trigger rules and revisable allocations

In 2013, the quarrel over north-east Atlantic mackerel had already made one point painfully clear: a fish stock can move faster than the bargain built to govern it. States kept their lines, fleets kept their habits, and the old allocation logic kept speaking in the language of yesterday's distribution. The fish did not listen. That was the deeper failure. A management system designed to preserve order had become a machine for replaying the same argument under worsening ecological conditions. The chart stayed still.

That is the right place to begin because trigger rules and revisable allocations are often misunderstood as clever technical refinements when they are really attempts to solve a more fundamental political problem. Fisheries bargains become brittle when every meaningful ecological shift requires a fresh diplomatic struggle. Renegotiation then ceases to be an occasional corrective and becomes the permanent climate of governance. States arrive at each season knowing that any change in distribution may reopen questions of fairness, access, and national standing. Fatigue sets in. So does opportunism. The value of trigger rules lies in the hope that some of this conflict can be moved out of improvisation and into design.

A trigger rule is a pre-agreed condition under which allocations, access terms, or management responses adjust once specified thresholds are crossed. The show is simple enough: if a stock spends more time in a certain zone, if survey evidence shows persistent distribution change, or if agreed indicators move beyond a defined range, then the management arrangement shifts according to terms already accepted in advance. The name for this is conditional adaptability. The explanation is that parties try to write future adjustment into the present bargain, so that not every ecological change must first become a diplomatic crisis. The implication is larger than administrative convenience. Flexibility becomes part of security. A system that can absorb movement without immediate political theatre is less likely to convert biology into confrontation.

This is why adaptability should be treated as a security feature rather than as a soft concession to uncertainty. A rigid system may look stronger because it speaks with the confidence of fixed rights and settled shares. In practice, rigidity often teaches actors to contest more aggressively once the ecological facts begin to drift away from the legal story. Fleet's fish to establish a presence. States issue unilateral quotas. Scientific disputes sharpen because so much now rides on them. The old bargain acquires symbolic weight precisely when it is least able to govern reality. Trigger rules attempt to interrupt that sequence. They say, in effect, that not every change in the fish requires a fresh struggle over who has yielded. Some changes can be anticipated institutionally and handled without public drama.

The attraction is obvious. So is the danger. Trigger rules reduce renegotiation fatigue only if the measurements that activate them are trusted. A threshold that exists on paper but is contested in practice does not absorb conflict. It redistributes it. Instead of fighting over shares directly, the parties now fight over whether the threshold has truly been crossed, whether the survey design is adequate, whether the season chosen for measurement is fair, and whether a temporary fluctuation is being dressed up as structural change. The rule becomes a second battlefield. The ruler matters.

This is why revisable allocations require more than legal drafting. They require an agreed politics of measurement. Who counts the fish? Over what period? In which part of the stock's life cycle? With what treatment of depth, seasonality, or anomalous years? Does the trigger activate automatically, or does it merely force review? Is the adjustment partial, capped, delayed, or immediate? None of these are technical footnotes. They determine whether the system is genuinely adaptive or merely inventive in the ways it can disappoint its participants. A state may accept revisability in principle and still resist a design that allows another state to claim too much from a single favourable reading of a moving stock. The issue is never only movement. It is a credible movement under the agreed interpretation.

The mackerel disputes in the north-east Atlantic remain the clearest illustration of what happens when such design is absent or weak. As the stock shifted and more of it became available in waters used by Iceland and the Faroe Islands, the established allocation bargain came under sustained pressure. The conflict did not arise simply because the fish moved. It intensified because no widely accepted mechanism existed to translate movement into adjustment without reopening the whole political question under conditions of mistrust. That is exactly the governance gap trigger rules are meant to close. Not by eliminating conflict, but by preventing each ecological change from demanding a theatrical renegotiation of status as well as catch.

A second example, more preventive than contentious, lies in the 2018 Agreement to Prevent Unregulated High Seas Fisheries in the Central Arctic Ocean. That agreement was not an allocation regime for an active commercial fishery, but it is still analytically valuable because it embodied the logic of anticipatory governance. The parties accepted that if the ecological and commercial situation changed in future, institutional response should not be improvised from scratch in a vacuum of trust. Consultation, science, and procedural structure were built in advance. This is the same governing instinct that lies behind trigger rules. Better to decide now how adjustment should begin than to discover later that every possible adjustment is already politically contaminated by first-mover anxiety.

This is where sovereignty enters the argument in its most sensitive form. States are often wary of revisable allocations because flexibility can sound like a disguised erosion of sovereign claim. A government may fear that agreeing to automatic or semi-automatic adjustment concedes too much to a future measurement process it does not fully control, or signals to domestic audiences that national rights are conditional in a way they would rather deny. That fear is real. It cannot be answered by procedural optimism alone. The deeper answer is that revisability need not mean surrender. A state can preserve its formal legal position while agreeing that operational shares or access terms will move within a bounded and pre-agreed framework if certain conditions are met. Sovereignty, in other words, need not require administrative immobility.

That distinction matters politically because it allows governments to say two things at once without necessarily deceiving anyone. They can insist that their underlying rights remain intact, while also agreeing that practical management must adapt to observed ecological change if the fishery is to remain governable. This is not always easy to sell domestically. Still, it is often easier to defend than an annual cycle of improvised bargaining in which every adjustment looks like capitulation because no one has accepted revisability in principle. Trigger rules can therefore lower the symbolic cost of adaptation by making it appear less like concession and more like the system working as designed. The mechanism takes some of the shame out of movement.

A useful limitation should be admitted. Not every stock, fishery, or political setting lends itself to clean trigger-based design. Some distributions are too seasonal, too noisy, or too scientifically contested for thresholds to carry enough authority. Some states may simply distrust one another too deeply to allow pre-agreed adjustments to activate without renewed political bargaining. Some trigger rules may create incentives for strategic fishing, selective measurement advocacy, or administrative delay. This is important because the adaptive argument is sometimes presented too elegantly. A poorly designed trigger can create a false sense of security while leaving the underlying distributive fight fully alive, merely postponed until the first contentious activation.

That limitation sharpens rather than weakens the case. The serious question is not whether every fishery should be governed by automatic formulas. It is whether systems can be designed to reduce the volume of political energy spent refighting predictable problems. A revisable allocation may need buffers, review periods, mixed indicators, caps on annual movement, or interim arrangements that distinguish between transient and persistent shifts. These are not signs of indecision. They are acknowledgements that ecological movement does not present itself in neat constitutional moments. It appears unevenly, seasonally, and often argumentatively. Adaptation has to be designed for that reality, not for a tidier fishery imagined in conference rooms.

This is also why trigger rules belong with the wider theme of conflict prevention. A regime that can revise itself under stress reduces the opportunities for brinkmanship. Fleets become less tempted to establish facts on the water if they know that distribution change will be handled through a recognised mechanism rather than only through unilateral pressure. States become less likely to overreact to each new survey result if the consequences of movement are already bounded and proceduralised. Scientific uncertainty remains. Political disagreement remains. But the path from ecological change to security tension grows less direct. That is a meaningful gain.

The quiet motif here is the ruler. It appears objective, almost innocent. Yet the ruler is never just a measuring device. It symbolises who gets to say when change is enough to matter and when it is still not enough to disturb the old settlement. A trusted ruler can calm a fishery because it allows movement to be recognised without immediate accusation. A distrusted ruler does the opposite. It turns every adjustment rule into another argument about bias, timing, and hidden advantage. The success of revisable allocations therefore depends not on flexibility alone, but on whether the parties believe the ruler itself belongs to the institution rather than to one side.

The broader lesson is that adaptive governance must be written before the pressure arrives. Once a stock has shifted visibly and political claims have hardened around that shift, it becomes much harder to persuade states that revisability is merely prudent administration. It looks instead like a demand to yield in the middle of an active dispute. Trigger rules work best when they are installed in calmer conditions, when parties can still think constitutionally rather than defensively. That is why their value is not only practical. It is temporal. They borrow calm from the present and spend it in a more difficult future.

The unresolved question, then, is not whether moving stocks requires flexible institutions. They do. The harder question is what kind of trigger a state can trust enough to let it alter real access without reading the alteration as a disguised loss of standing - and whether that trust can be built before the next shift in distribution turns the ruler itself into the most contested object in the fishery.

Scenario planning and institutional readiness

In October 2018, when the parties to the Central Arctic Ocean fisheries agreement chose restraint over premature exploitation, they were doing more than expressing caution about fish. They were acknowledging a harder institutional truth: the real danger in a new fishery often lies not in ignorance alone, but in what agencies do when pressure arrives before roles, information flows, and decision authority are ready. The first test is rarely scientific. It is organisational. A plan on paper is not enough.

That distinction matters because fisheries governance still carries a persistent temptation to confuse prediction with preparedness. Governments ask where the fish will move, which fleets will arrive, when conflict may intensify, or whether co-operation will hold. Those are sensible questions. They are not the most useful ones. Scenario planning matters less because it predicts the future accurately than because it forces institutions to confront what they would actually do if several plausible futures arrived at once. Who decides? Who informs whom? Which threshold triggers operational action? Which agency owns the public message? A chart can forecast movement. It cannot assign responsibility.

This is why scenario work should be understood as an exercise in institutional alignment rather than speculative storytelling. The show is practical enough: an agency imagines an enforcement surge around a newly important stock, a sudden arrival of new entrants, a breakdown in a joint monitoring arrangement, or an abrupt rise in contested inspections. Officials then trace what would happen next. The name for this is readiness planning. The explanation is simple. Most failures during fast-moving fisheries pressure come not from a total absence of authority, but from confusion over sequence - who has the mandate, who holds the data, who can order patrol redeployment, who can speak publicly, and who can approve an interim measure before the season escapes. The implication is that preparedness is visible in choreography, not prophecy.

This matters especially in moving-stock fisheries because ecological change compresses political time. A stock may become newly valuable in waters not previously central to management. A fleet may redirect effort quickly if it senses opportunity. A regional body may still be debating the implications while vessels are already adjusting in practice. Under such conditions, the weakness of a state is rarely that it lacks any rule at all. More often, its weakness lies in the interval between recognition and action. Scenario work narrows that interval by exposing ambiguity before an incident forces everyone to interpret it under pressure.

A useful example lies in the logic of the 2018 Central Arctic Ocean agreement itself. The agreement did not rest on confidence that every future development could be foreseen. Its value lay in creating some institutional preparedness before commercial pressure fully matured. Consultation channels, scientific co-ordination, and a recognised framework for future decision-making were all, in effect, elements of scenario readiness. The parties accepted that uncertainty would remain. What they sought to reduce was the chance that uncertainty would meet administrative improvisation.

This is often the most realistic ambition available in Arctic governance. One cannot script the fish. One can script the first institutional moves.

A second example can be read from the recurrent tensions over north-east Atlantic pelagic stocks such as mackerel. Those disputes revealed how quickly stock movement, unilateral quotas, scientific disagreement, and domestic politics can compound one another when no shared anticipatory structure exists to handle an altered distribution. The parties did not lack intelligence. They lacked a sufficiently trusted and prepared mechanism for converting emerging ecological change into coordinated administrative response before bargaining hardened. Scenario planning could not have removed the distributive conflict. It might, however, have clarified which agencies needed to be ready for enforcement surges, diplomatic messaging, and temporary management responses once the conflict began to materialise.

This is where the practical value of scenario work becomes clearest. Good scenarios do not ask, "What will happen?" in the abstract. They ask, "If this happened next month, who would do what by the end of the day?" That question changes the quality of preparation. An enforcement surge is no longer a theoretical concern; it becomes a staffing and patrol problem. A new entrant is no longer just a political irritation; it becomes a licensing, monitoring, and communication problem. A sudden collapse in co-operation is no longer a lamentable geopolitical mood; it becomes a test of whether national agencies know how to replace lost information flows, reassign tasks, and manage public signalling without escalating by accident. Preparedness begins when institutions can answer these questions without looking around the room in silence.

That silence is often what scenario work is designed to prevent. Fisheries administrations are typically spread across ministries, coast guards, research bodies, port authorities, prosecutors, and political offices, each holding part of the relevant capacity. Under normal conditions, fragmentation may be manageable. Under stress, it becomes dangerous. A scenario exercise can expose whether data held by one office will actually reach the decision-maker who needs it in time, whether a patrol request will wait for legal clearance that has not been pre-arranged, whether public statements will be delayed because no communications authority was agreed, or whether an interim diplomatic contact will be overlooked because each agency assumed another was responsible. These are not glamorous problems. They are usually the ones that matter first.

A sensible limitation should be admitted. Scenario planning is not a magic shield against political conflict, and it can fail in familiar ways. Exercises may become ritualised, too generic, or too detached from the real frictions of budget, hierarchy, and inter-agency rivalry. Officials may produce elegant documents that no one consults when pressure arrives. Governments may also plan for the wrong contingencies because the scenarios are shaped by what is comfortable to discuss rather than by what is institutionally awkward. This is important because preparedness cannot be measured by the existence of a plan alone. It must be tested against whether people know their roles and trust the pathways through which decisions will move.

That limitation, however, does not reduce the need for scenario work. It sharpens the standard by which it should be judged. A good scenario process is one that reveals weak information flows, unclear authority, and missing legal or operational triggers early enough to fix them. It also distinguishes between planning and prediction. The point is not to say that a particular enforcement surge or diplomatic rupture will occur on a certain timetable. The point is to prevent plausible stress from becoming paralysing novelty. A government does not need to foresee the exact form of the next dispute to know that it should have already settled who calls whom, who can approve a boarding posture, and who can convene a technical review within hours rather than days.

This is why institutional readiness should be treated as a practical lens for assessing governance capacity. One can ask simple questions. Does an agency know what counts as an operational trigger? Can it shift from routine monitoring to surge mode without improvised bargaining over mandate? Are responsibilities between fisheries, foreign affairs, coast guard, and legal offices understood in advance? Can public communication remain credible if co-operative channels narrow suddenly? These questions are more revealing than abstract claims about resilience. They test whether the state has converted awareness of uncertainty into a usable sequence.

The quiet motif here is the folder on the desk. It may contain maps, contact lists, authority matrices, draft statements, patrol options, and legal notes. It looks unimpressive. Under stress, it can be the difference between orderly action and institutional drift. The folder matters not because it predicts the future, but because it prevents the first crucial hours from being consumed by guessing who owns the problem.

This also helps explain why scenario work belongs to conflict prevention rather than merely to administrative prudence. A fisheries dispute intensifies quickly when agencies send mixed signals, delay action because authority is unclear, or improvise public language that accidentally raises the stakes. Prepared institutions do not avoid pressure, but they make it less likely that pressure will be amplified by internal confusion. That is a real security good. It keeps more disagreements in the category of governance problems and fewer in the category of public crises.

The deeper lesson is that readiness is a relational property. It lives in the fit between agencies, between data and decision, between law and operations, between central government and local knowledge, and between ordinary routines and surge conditions. Scenario planning is valuable because it reveals where that fit is weak before fleets, media, and rival states discover the weakness first. Preparedness is therefore less about imagination than about administrative honesty.

The unresolved question, then, is not whether moving stocks and fragile cooperation justify planning. They clearly do. The harder question is what kind of institution can rehearse enough to act coherently under stress without becoming so attached to its scenarios that it mistakes readiness for control - and whether the next real test will expose a government that has prepared for pressure, or merely described it elegantly in advance.

Building compliance into incentives

In 2009, the Food and Agriculture Organisation adopted the Port State Measures Agreement, and a simple governing insight moved closer to the centre of fisheries policy: a fishery is easier to police when lawful behaviour pays more reliably than evasive behaviour. That may sound obvious. It is not the way many systems are built. Too often, regulators ask surveillance and sanctions to correct incentive structures that were poorly designed from the start. The hold fills. The lesson spreads.

That is the right place to begin because compliance is often discussed in moral language when it is, more stubbornly, a matter of comparative advantage. A vessel operator deciding whether to report fully, accept stricter access conditions, or stay within the spirit as well as the letter of the rules is not responding only to civic duty. The operator is responding to timing, margins, access to ports, fee structures, quota arrangements, the likelihood of inspection, the ease of laundering catch, and the commercial value of a clean paper trail. Where those elements reward opacity, grey tactics thrive. Where they reward credible legality, enforcement becomes less lonely.

This is why building compliance into incentives matters so much for moving-stock fisheries. The show is practical: a state ties access to reporting quality, conditions entry on verifiable documentation, links continued participation to audit performance, or shares benefits in ways that make lawful co-operation worth more than short-term brinkmanship. The name for this is incentive design. The explanation is simple. Behaviour changes more durably when rules alter the payoff structure than when authorities rely only on detection and punishment after the fact. The implication is substantial. A fishery governed through better incentives needs fewer dramatic interventions to produce the same, or better, operational outcomes.

Fees are one of the clearest and most misunderstood instruments in this space. A fee is not merely a revenue device. It can be used to sort conduct. If access fees, monitoring charges, or landing costs are structured clumsily, they may push operators towards opacity by making full compliance too expensive relative to evasion. If structured intelligently, they can reward verified behaviour and raise the cost of ambiguous practice. The point is not to fine one side and flatter the other. It is to make documented, lawful participation commercially steadier than opportunistic manoeuvre. A fee schedule can therefore act like a quiet enforcement map, guiding operators towards the behaviour the state would otherwise have to compel more noisily.

The logic of port state control after 2009 offers one important example. States using port access, service conditions, and documentation requirements do more than inspect fish that have already been caught. They change the calculation before the trip begins. A vessel that knows doubtful catch may face denial of landing, delay, or costly scrutiny; it must factor that risk into every earlier decision: where to fish, how to report, whether to tranship, whether to manipulate identity, whether to operate at the edge of a contested zone. The policy instrument acts upstream. That is why access conditions matter so much. They turn routine commerce into a system of conditional permission, and conditional permission can be a stronger deterrent than theatrical punishment.

A second example lies in quota and access bargains where benefit-sharing is used to stabilise compliance. This can take different forms: side payments, community-linked allocations, access swaps, or arrangements that allow states and fleets to retain enough value from co-operation that unilateralism looks less attractive. The precise design varies, but the governing idea remains consistent. If actors believe the gains from staying inside the arrangement are tangible and reasonably distributed, they are less likely to treat grey tactics as a necessary correction to an unfair bargain. Benefit-sharing does not abolish conflict. It reduces the number of cases in which conflict appears to be the only profitable language left.

This is where political economy enters the enforcement story in its most practical form. A patrol vessel can board, a prosecutor can fine, and a ministry can protest, yet none of these will settle the daily behavioural question if the underlying payoffs still reward risk. The abstraction here is compliance culture. Its visible form is a choice on deck or in an office: whether to file fully, whether to seek legal access rather than opportunistic entry, whether to maintain tamper-evident systems, whether to accept a lower immediate gain in exchange for continued participation in a higher-trust regime. The explanation is that operators learn from patterns in cost and reward far more quickly than they learn from slogans about responsibility. The implication is uncomfortable for some regulators. If grey conduct remains profitable, enforcement alone is being asked to perform too much of the governance burden.

Access conditions are especially useful because they connect permission directly to behaviour without requiring maximalist sanctions at every breach. A state can insist on verified reporting, functioning monitoring equipment, timely documentation, or acceptance of specific audit protocols as conditions of access rather than as after-the-fact obligations only. That distinction matters. It changes the operator's posture from calculating how much non-compliance might survive scrutiny to calculating what must be done to remain inside the fishery on acceptable terms. A licence then becomes not only a formal authorisation, but a behavioural contract. The rule is no longer waiting passively for breach. It is actively shaping entry.

Still, one should resist the temptation to treat incentive design as a technocratic cure-all. Poorly designed incentives can damage legitimacy as easily as they improve compliance. Fees can burden smaller lawful operators more heavily than larger evasive ones. Benefit-sharing can be captured by political insiders. Access conditions can become so administratively dense that compliance itself turns into a privilege of capital and paperwork fluency rather than a general expectation. This is a serious limitation. A fishery in which lawful participation is too expensive or too procedurally burdensome may not become more honest. It may simply become more concentrated, more resentful, or more dependent on intermediaries who can navigate the bureaucracy.

That limitation matters because behaviour is shaped not only by incentives in the narrow economic sense, but by whether those incentives appear fair enough to command obedience. A state that designs clever payoffs while ignoring distributional justice may still reduce one form of grey conduct and generate another. Fishers priced out of direct access may rely more heavily on leasing, informal arrangements, or opaque partnerships. Communities excluded from the gains of co-operation may pressure governments to bargain more aggressively abroad. Compliance then weakens politically even if the instruments looked elegant on paper. Incentive design, to work, must be socially as well as economically legible.

A useful counterpoint should also be admitted. Not every breach or evasive tactic is best answered by altering the payoff structure. Some conduct deserves direct sanction because it threatens the credibility of the whole regime. A system built entirely on inducement may invite testing if operators conclude that every difficulty will be answered with another adjustment to keep them inside the arrangement. This is why incentive design and enforcement are complements, not substitutes. The serious aim is not to remove punishment from fisheries governance. It is to ensure punishment is reserved for conduct that remains irrational or unacceptable even after the ordinary incentives have been arranged to favour compliance.

This is where moving stocks make the problem sharper. When fish shift across zones or into newly important areas, old bargains often leave some actors feeling stranded. If the adaptive system offers no credible path to lawful gain, grey tactics begin to look like rough self-help. A vessel entering a contested zone, stretching reporting rules, or pressing at a boundary may not believe it is being reckless. It may believe the legal system has failed to keep pace with ecological change. Whether that belief is correct matters less, in behavioural terms, than whether the legal system can answer it with incentives that make co-operation preferable to improvisation. Adaptation that ignores payoffs will often end up chasing its own failures with more inspection.

The quiet motif here is the ledger. Not the moral ledger of virtue and vice, but the ordinary ledger of costs, shares, conditions, and expected returns. Fishers, firms, and states all keep one, whether formally or not. A governance system that wants compliance must change that ledger in visible ways. It must make the lawful route steadier, more predictable, and more worthwhile than the shadow route, without pretending that everyone begins from the same position or responds to the same pressure. If the ledger continues to favour ambiguity, then ambiguity will keep finding crews, lawyers, and markets ready to carry it.

The broader lesson is that operational outcomes are written partly in policy design long before they appear in patrol logs or court files. A boarding, a seizure, or a port denial is often the final expression of an incentive problem the state noticed too late. Better design cannot remove all conflict, but it can reduce the number of occasions on which ordinary opportunism becomes strategically sensible. That is already a major gain. It lowers the temperature of enforcement by shifting more behaviour towards the lawful route before the vessel ever leaves harbour.

The unresolved question, then, is not whether fees, access conditions, and benefit-sharing can influence conduct. They plainly can. The harder question is what kind of incentive system can make compliance the rational choice without turning lawful participation into an expensive administrative obstacle course - and whether governments are willing to judge their policies not by how sternly they speak about rule-breaking, but by how effectively they have made rule-following the better bargain in the first place.

Technology governance and verification norms

In 2018, when the parties to the Central Arctic Ocean fisheries agreement chose to build consultation and scientific co-operation before a commercial fishery had properly taken shape, they were doing something larger than precaution. They were acknowledging that information itself would have to be governed before it could stabilise anything. A satellite image, a vessel track, a camera record, a shared database - none of these can calm a dispute merely by existing. A file can still be doubted.

That is the right place to begin because fisheries debates too often talk about technology as though more tools naturally produce more order. Sometimes they do. More often they produce more evidence, more points of comparison, and, if governance is thin, more arguments about what the evidence means, who controls it, how it travelled, and whether it has been manipulated. A new sensor can narrow the search problem while widening the interpretive one. An electronic record can look authoritative and still fail to persuade if its chain of custody is weak or its standards are not shared. The machine sees. The quarrel remains.

This is why tools matter less than the norms surrounding them. The show is familiar enough: one authority relies on vessel tracking data, another questions the integrity of the signal; one inspection regime trusts camera footage, another disputes tamper resistance or review protocol; one state shares a dataset, another asks who validated it, when it was cleaned, and whether the format omitted relevant context. The name for this is verification governance. The explanation is straightforward. Technology enters fisheries management as evidence only when institutions agree, at least broadly, on how records are generated, preserved, compared, and challenged. The implication is substantial. Without shared norms, innovation can become just another arena of dispute rather than a route out of one.

This matters especially in moving-stock fisheries because ecological change already strains trust. When fish distributions are shifting, quota bargains are under pressure, and enforcement choices are politically loaded, every new technical input risks being read strategically. A tracking gap may be treated as concealment by one side and ordinary signal instability by another. A camera system may be praised as transparent and denounced as intrusive. A shared database may look like co-operation from one desk and selective disclosure from another. Under those conditions, the decisive question is not simply what technology can do. It is what the parties have agreed counts as valid use, reliable handling, and acceptable challenge. Norms decide whether data travel.

Standards for data sharing are among the most important of these norms because they determine whether institutions are working with the same object or merely with differently cleaned and differently timed versions of it. The show here is quiet but decisive: common formats, agreed submission windows, known metadata requirements, documented revisions, and stable expectations about who may access what and when. The explanation is that evidence weakens when each institution rebuilds the same dataset through its own conventions and then treats discrepancy as proof of bad faith. The implication is why shared standards calm a fishery. They do not eliminate disagreement over meaning, but they reduce avoidable disagreement over whether the underlying record was even comparable.

Tamper evidence matters for a similar reason. In fisheries governance, many technologies are not trusted because they are impossible to manipulate. They are trusted because manipulation leaves signs that can be checked. A camera housing that shows interference, a monitoring unit that records interruption, a transmission system that logs anomalies, or an audit trail that reveals who accessed and altered data does not produce purity. It produces contestable integrity. That distinction is critical. Authorities and operators do not need to believe a system is metaphysically incorruptible. They need to believe that interference is discoverable enough to make denial expensive. The name for this is visible integrity. The explanation is that technology becomes governable when the system can show not only what it recorded, but whether the record itself has been handled in a defensible way. The implication is powerful. Trust often rests less on prevention than on traceability of breach.

Auditability extends this logic across institutions. A record that cannot be independently checked is rarely stabilising for long, especially where the underlying political stakes are already high. Auditability means more than opening files to scrutiny. It means building systems in which another competent body can reconstruct what happened: what data were collected, through which device, under whose authority, stored in what form, amended or accessed by whom, and linked to what later decision. The show is procedural rather than dramatic. The explanation is that fisheries disputes are often won or lost in administrative reconstruction, not in the first excitement of detection. The implication is why verification norms must be portable. A record that convinces one agency but collapses before another has not really stabilised the system.

A useful example lies in the wider spread of port state controls and electronic documentation regimes after 2009. The real strength of such systems has never been simply that they digitise information. Their force lies in the growing expectation that documentation must be comparable, reviewable, and sufficiently auditable to support denial of landing, service restriction, or further inquiry without constant reinvention of evidentiary standards. Where those expectations are strong, technology supports governance. Where they are weak, digital paperwork can become little more than a faster route to the same old arguments about origin, timing, and responsibility.

A second example can be read from the recurrent difficulties around vessel tracking, spoofing, and identity management in remote fisheries. The lesson there is not that tracking technology has failed. It is that its stabilising value depends on norms about verification strong enough to interpret anomalies without either naivety or panic. A suspicious signal jump, identity mismatch, or transmission gap does not speak for itself. It becomes usable when institutions share enough standards to distinguish likely malfunction from likely manipulation, and when those standards are accepted widely enough that enforcement or diplomatic follow-up does not begin by relitigating the meaning of the technology itself. Otherwise the signal becomes part of the dispute rather than evidence within it.

A sensible limitation should be admitted. Shared norms do not abolish strategic behaviour. Actors will still search for blind spots, exploit timing gaps, challenge unwelcome records, or invest in new countermeasures once older ones become too transparent to be profitable. This matters because technology governance should not be romanticised into a final settlement between truth and deception. It is a moving contest. The serious ambition is not to design a system that cannot be tested. It is to design one in which testing becomes harder, more detectable, and less likely to destabilise the whole evidentiary environment each time it occurs.

That is why governance must anticipate countermeasures rather than merely celebrate innovation. Every new monitoring or documentation tool invites response. Operators may interfere physically, manipulate identities, pressure human reviewers, exploit formatting differences, or shift behaviour into the spaces where the tool is least authoritative. Institutions must therefore ask not only whether a technology works under normal conditions, but how it is likely to be challenged once the users understand its strengths. The policy error is to treat adoption as the end of governance. Adoption is the beginning of a new round of adaptation. A seal gets tested.

This anticipation is what allows verification norms to travel across institutions. When a tamper-evident culture, a disciplined audit practice, and a shared expectation of comparable data formats are established in one part of the system, they can begin to shape adjacent practices: port inspections, licensing checks, shared science, electronic monitoring review, financial due diligence, and even diplomatic exchanges over incidents at sea. The technology itself may differ from office to office. The norm remains recognisable. That continuity is what makes governance stronger than gadgetry. It allows different institutions to treat evidence in related ways even when their mandates diverge.

The deeper political point is that norms can institutionalise restraint. A state that trusts the audit trail of a shared record may feel less need to dramatise an incident publicly. A fleet that knows tampering will likely be discovered may find grey tactics less attractive. A prosecutor that can rely on consistent data handling standards may pursue cases more confidently. A joint commission may argue over allocation while still accepting the integrity of certain categories of evidence. These are not minor gains. They reduce the number of disputes in which technology itself becomes the first contested object. That is already a form of conflict prevention.

The quiet motif here is the seal. A seal on a device, a file, a transfer log, an access record. It seems small. It matters because it symbolises a larger institutional promise: not that no one will ever try to interfere, but that interference will leave marks the system is organised to notice. Fisheries governance needs more of that promise than it needs breathless faith in the next tool. A seal can carry more peace than a sensor if the seal belongs to a norm everyone recognises.

The harder question, then, is not whether better tools will continue to arrive. They will. The sharper question is whether institutions can agree on standards of sharing, tamper evidence, and auditability strong enough that the next disputed signal, camera record, or database entry is treated as evidence to be tested within a common framework rather than as the opening shot in yet another argument about the machinery itself.

Chapter 13

A Framework for Fisheries Conflict

How does a fishery become dangerous without becoming dramatic first? That is the question left standing after every disputed boarding, every strained quota bargain, every season in which the fish move a little faster than the institutions meant to contain the consequences. Conflict in fisheries rarely begins as spectacle. It begins as pressure accumulating in ordinary places: in an agency that cannot process the file quickly enough, in a fleet that sees a better return in ambiguity than in compliance, in a monitoring system that can detect movement but not settle attribution. The net comes up. So does the strain.

That is why a useful framework matters. Not because it offers the comfort of a master theory, and certainly not because it can predict every incident in advance, but because fisheries disputes are too often described in fragments. One argument focuses on law, another on science, another on nationalism, another on surveillance, another on the behaviour of vessels at sea. Each may be right. None is sufficient alone. What is needed is a way of comparing very different fisheries without flattening them into sameness, and of recognising that what appears episodic from a distance is often structural when viewed through the right lens.

The central tension is simple enough to state and hard to govern. Stocks move through environments where enforcement remains uneven, incentives remain sharp, and visibility remains partial. States respond with rules, patrols, and bargains that may look solid until ecological change, market pressure, or political mistrust exposes the gaps between what can be proclaimed, what can be enforced, and what can be proved. Conflict risk therefore does not sit mainly in dramatic incidents. It sits in the relationship between capacity, incentives, and visibility, and in the way those three conditions magnify one another when stress arrives. The frame matters because the relationships matter.

This is a harder way of seeing than the familiar language of sudden crisis. It asks the reader to treat a fisheries dispute not as an isolated event but as a pattern produced by underlying arrangements. Capacity is not merely a coast guard budget or a legal competence on paper. It is the actual ability to administer, verify, sanction, and follow through. Incentives are not merely greed or bad faith. They are the payoffs built into quota structures, market channels, access conditions, and domestic politics. Visibility is not simply a technical question about sensors. It is the broader problem of what can be seen, what can be attributed, and what can be turned into durable evidence. A checklist, used properly, disciplines thought. It stops a quarrel from hiding behind its own preferred explanation.

The stakes reach beyond analytical tidiness. A government that misreads structural strain as a series of disconnected incidents will tend to respond theatrically and late. It will send a patrol where a better sanction pipeline was needed, tighten rhetoric where incentive design was doing the real damage, or buy another tool where the missing ingredient was trust in the norms that govern verification. A government that reads the structure more clearly may still face conflict, but it has a better chance of acting proportionately before the dispute begins to define the fishery itself. The difference is not academic. It shapes whether a frontier is lived as routine friction or recurring emergency.

This matters later as much as now because moving-stock fisheries reward learning on all sides. Fleets learn which rules are real, which boundaries are thinly policed, which ambiguities are profitable, and which governments are slower than they sound. Communities learn whether institutions absorb change fairly or merely reallocate pain. States learn whether old bargains can still carry ecological movement without political rupture. If those lessons accumulate in the wrong direction, stability begins to look like a nostalgic word for a fishery already operating on borrowed time. The point of a framework is not to promise certainty. It is to stop wishful thinking from posing as analysis.

The path ahead tightens the book's argument into a more portable form by bringing the core triad into one working frame and then showing where strain usually gathers before anyone is ready to call it crisis. From there the inquiry turns towards the conditions of a frontier that remains contested yet governable: not fixed in place, not purified of disagreement, but organised well enough that change does not automatically become conflict. What emerges is less a blueprint than a discipline of attention, a way of asking better questions before the next season, the next disputed inspection, or the next shift in distribution begins to harden into something larger.

A useful caution belongs here. No framework, however well designed, will abolish political choice or dissolve the asymmetries of power that shape distant-water governance. Some states will still prefer ambiguity, some fleets will still press at the margin, and some institutions will still fail under pressure despite careful design. The value of the framework lies elsewhere. It helps distinguish what is feasible from what is fantasy. It refuses the comforting fiction that permanent agreement, frozen borders, or complete visibility are realistic foundations for stability in moving seas. Better governance begins when those fantasies are set aside.

The harder question, then, is not whether the frontier can be made calm. It cannot, not in any final sense. The sharper question is what kind of fishery becomes more dangerous when one part of the triad weakens first: the one with thin enforcement, the one with stronger incentives for evasion, or the one in which visibility fails just enough that every other strain can hide inside it until the next ordinary incident reveals how much was already building beneath the surface.

Capacity, incentives, and visibility

In January 2019, the dispute over the Russian trawler *Melkart-5* in waters around Svalbard looked, at first glance, like a single difficult incident: an inspection, a refusal, an escort, a protest. It was all of those things. It was also something more revealing. The event exposed several different weaknesses at once - how much authority could be operationalised, how much pressure existed to resist or delay, and how far the available evidence could travel before it dissolved into argument. One vessel. Three questions.

That is the right way to begin because fisheries conflict is too often narrated in the grammar of episodes. A season turns tense, a patrol boards, a state protests, a fleet pushes too hard, a headline appears, and analysts begin explaining what "caused" the incident. Yet incidents are rarely self-explanatory. They sit inside structures that make certain kinds of friction more likely, more profitable, or harder to settle once they appear. A contested boarding in a well-administered, low-incentive, high-visibility fishery means something very different from a similar boarding in a fishery where patrols are thin, quotas highly valuable, and attribution uncertain. The surface action may resemble itself. The governing context does not.

This is why the triad of capacity, incentives, and visibility matters so much. Capacity asks what the state and its institutions can actually do, not merely what they are legally entitled to do. Incentives ask what actors gain by pressing, delaying, concealing, or co-operating. Visibility asks what can be seen, what can be proved, and what can be attributed strongly enough that a record becomes a sanction rather than an argument. The point of the triad is not conceptual elegance. It is practical comparison. It gives a way to place very different fisheries on common dimensions without pretending they are otherwise identical. Once those dimensions are clear, conflict begins to look less like weather and more like architecture.

Capacity is usually the easiest element to romanticise and the hardest to measure honestly. States like to point to vessels, aircraft, coast guards, laws, agreements, and monitoring systems. Those things matter. But capacity only becomes real when they function together as an administrative pipeline. The show is concrete: a patrol can reach the scene, a boarding can occur safely, a report can be written properly, evidence can be stored and compared, a legal threshold can be met, and a sanction can survive review. The name for this is operational and administrative capacity. The explanation is that enforcement does not happen at the horizon alone; it continues through offices, courts, licensing systems, and information channels. The implication is why two fisheries with similar laws can behave so differently. One has paper capacity as well as patrol capacity. The other has declarations.

The *Melkart-5* case showed precisely this. The issue was never just whether an inspection could be justified. It was whether justification could be carried through politically, operationally, and evidentially once resistance began. Norway's comparatively strong institutional machinery made a crucial difference there. That strength did not remove friction or prevent diplomatic contest. It did mean the incident unfolded inside a state apparatus capable of linking observation, action, and follow-through more tightly than many jurisdictions could. Capacity, in other words, is best judged not by whether conflict occurs, but by how much organised response survives once conflict has started.

A second example lies in the broader experience of the north-east Atlantic mackerel dispute after stock movement unsettled earlier bargains. The conflict is often read mainly as a dispute over allocation. It was that, but capacity shaped it too. Different states were not equally positioned to monitor effort, absorb diplomatic strain, defend unilateral decisions administratively, or sustain prolonged bargaining without operational slippage. The same ecological change entered different institutional environments and therefore produced different risks. This is exactly what the triad helps reveal. Capacity is not background. It is one of the conditions under which fish movement becomes politically combustible or administratively manageable.

Incentives bring the second dimension into view and often explain more behaviour than moral language does. A vessel, company, or state does not need to be villainous to generate conflict. It may simply be responding to a payoff structure that rewards assertion more than restraint. High quota value, squeezed seasonal windows, weakly designed access terms, reputationally tolerable opacity, and domestic political gain from firmness can all make confrontation or brinkmanship look rational. The show is visible enough: more aggressive boundary fishing, slower compliance, pressure for unilateral shares, resistance to inconvenient science, or greater appetite for ambiguous conduct. The explanation is not always greed in the crude sense. It is often comparative gain. If the expected return from pushing exceeds the expected cost of being checked, pushing becomes ordinary.

This is why incentives should be treated as political economy rather than psychology. A fishery with thin profits, heavy debt, contested shares, and easy routes into secondary markets will generate different conduct from a fishery in which access depends on clean documentation, benefit-sharing is credible, and market consequences for doubtful behaviour are immediate. The same operator may behave differently under each regime. The implication is unavoidable. Conflict risk sits partly in the design of the payoffs. Where grey tactics, unilateral claims, or delay are commercially or politically rewarding, enforcement ends up trying to correct a deeper design problem. That is a losing arrangement over time.

Visibility is the third dimension, and perhaps the easiest to misunderstand because modern fisheries discourse is saturated with talk of technology. Satellite imagery, AIS, VMS, electronic monitoring, databases, and digital documentation can all widen what authorities know. Yet visibility is not the same as data abundance. A fishery can be rich in traces and poor in usable evidence. The show is familiar: a suspicious track appears, a camera records something partial, a logbook entry conflicts with a landing, a vessel identity is blurred, or a sensor gap invites competing explanations. The name for this is contested visibility. The explanation is that governance depends not only on seeing conduct, but on turning that sight into proof, attribution, and shared meaning. The implication is why conflict risk rises when monitoring outpaces verification norms. More can be seen. Not enough can be settled.

The value of visibility as an analytic dimension becomes clearest when comparing fisheries that look similar in legal or ecological terms but differ in evidentiary environment. One fishery may have limited patrol reach but strong port documentation, clean vessel identity rules, and credible audit trails. Another may possess sophisticated sensors and still struggle because spoofing, fragmented data, or weak evidentiary chains make every case arguable. The first may prove more governable in practice. Visibility, then, is not a trophy for technological sophistication. It is a question about whether institutions can agree on what the record means strongly enough to act on it.

A useful limitation should be admitted. The triad does not explain everything. National identity, historical grievance, strategic culture, and sheer bad luck still matter. A fisheries dispute can be intensified by symbolism or domestic politics in ways no analytic checklist fully captures. That matters because a framework can become a crutch if it encourages false completeness. Yet the limitation is less damaging than it sounds. The triad is not meant to replace history or judgement. It is meant to discipline them. It asks better questions before analysis begins to drift into stories that over-credit personality, legal phrasing, or one dramatic encounter while ignoring the structural conditions that made the encounter meaningful.

This is where the checklist idea becomes most useful. When comparing two fisheries, one can ask: how much usable capacity exists from detection to sanction? What incentives are actors facing if they comply, stretch, or conceal? How visible is conduct in the forms that matter for attribution and follow-through? These are not abstract prompts. They direct attention to observable features: the patrol-to-paper pipeline, the profitability of ambiguity, the strength of audit trails, the political reward for firmness, the market cost of doubtful documentation, the administrative ability to move a file before it dies. Once those features are examined together, a fishery begins to reveal whether its conflict risk is episodic noise or structural strain.

The deeper claim follows from that. Conflict risk is structural, not episodic, because the same patterns recur wherever weak capacity, strong incentives to press, and limited or contested visibility overlap. The particular stock, flag, or incident may change. The architecture does not. A sudden distribution shift matters more where administrative systems are thin. A compressed season matters more where quota values are high. A disputed inspection matters more where evidence chains are fragile. The triad helps explain why the same ecological trigger can pass quietly in one fishery and ignite sustained tension in another. It also clarifies why reforms that target only one dimension often disappoint. Better tools cannot compensate fully for bad incentives. Harsher sanctions cannot compensate fully for poor visibility. More patrols cannot compensate fully for administrative weakness.

The quiet motif here is the file. Capacity determines whether the file moves. Incentives shape why the file exists in the first place. Visibility decides whether the file contains something durable or merely suggestive. Seen this way, a fisheries dispute is not just a confrontation on the water. It is a structure made briefly visible. One vessel, three questions.

The harder question, then, is not whether a fishery has experienced incidents before. Almost all contested fisheries have. The sharper question is what those incidents were sitting on top of: a system with enough capacity, fair enough incentives, and clear enough visibility that conflict remained containable, or a system where one of the three was already weak enough that every minor dispute was really a signal of deeper structural strain waiting for the next season to expose it more fully.

Early warning indicators and stress points

In 2013, the dispute over north-east Atlantic mackerel did not begin with a single explosive act. It accumulated through altered stock presence, hardening quota claims, scientific disagreement, and the gradual erosion of confidence that the old bargain could still absorb the new distribution. By the time the conflict was widely recognised as serious, the warning signs had already been visible for some time. The pressure had been building in plain view. The water changed first.

That is the right place to begin because fisheries conflict rarely arrives as a lightning strike. More often, it comes as a sequence of small frictions that remain politically manageable until they no longer do. A state notices more fish in waters that previously mattered less. A season becomes shorter and more crowded. Scientists disagree not only about what they know, but about what should count as enough certainty to change policy. Enforcement agencies detect more anomalies than they can turn into credible sanctions. None of these developments, taken alone, compels a crisis. Together they can produce a system already under strain before anyone chooses to call it unstable.

This is why early warning should be understood not as prophecy but as disciplined attention to stress points. The show is familiar enough: a stock shifts abruptly or more persistently than institutions expected; the fishing window tightens and fleets crowd into the same period or area; scientific advice becomes more publicly contested; sanction pipelines begin to lag behind the number or complexity of cases. The name for this is accumulative strain. The explanation is that governance systems usually fail through the interaction of pressures rather than through one dramatic rupture. The implication is crucial. Minor disputes are often signals, not noise.

Sudden distribution shifts are among the clearest of these signals because they expose the gap between ecological movement and institutional speed. A stock appears more regularly in a zone where its presence was once marginal, or leaves a zone whose political and economic expectations were built around historical abundance. The biological event matters, of course. But its real significance lies in what it does to allocation logic. Quota shares that once looked settled become newly arguable. Surveillance plans designed for older patterns begin to miss where effort is now concentrating. Governments that had treated revision as a distant question are forced into improvised language about fairness, entitlement, and scientific caution. The fish move. The file thickens.

The mackerel dispute remains an obvious example because it showed how distributional change can become politically dangerous when no trusted mechanism exists to translate altered presence into orderly adjustment. The fish did not merely cross an invisible line and trigger a legal puzzle. They unsettled the credibility of an existing bargain. That is the key early warning. When a new distribution begins to make an old allocation look less like order and more like inertia, the system has entered a zone of mounting risk even before the first patrol or ministerial statement intensifies matters.

Compressed seasons create a different but equally important stress point. A short, intense fishing window magnifies every existing weakness. Patrol assets are stretched because more effort is concentrated into fewer days. Port inspections become more difficult because landings and paperwork surge at once. Fleets feel stronger incentives to race, delay, or take advantage of ambiguity because the commercial opportunity is fleeting. The show is visible: crowded grounds, sharper competition, more hurried reporting, more pressure on officers to decide quickly, less time to resolve a doubtful event before the economic value of the trip has already been realised. The explanation is simple. Compression rewards speed and punishes administrative delay. The implication is why systems that function tolerably under average conditions may fail under seasonal pressure.

A useful second example lies in the recurrent politics of mixed-stock and pulse fisheries in northern waters, where bycatch constraints, target-stock movement, and short openings can create a situation in which enforcement demand rises exactly when verification becomes hardest. Authorities may know more fishing effort is converging. They may even anticipate the pressure. Yet if roles, staffing, and sanction pathways are not aligned for that compressed period, minor breaches or contested interactions begin to multiply. This does not necessarily mean the fishery has entered crisis. It does mean that stress is moving from structural background to operational foreground.

Contested science is another early warning indicator, not because disagreement among scientists is itself pathological, but because certain kinds of disagreement become politically charged in predictable ways. A healthy science process includes uncertainty, revision, and argument. Trouble begins when those arguments stop being processed mainly as normal intellectual labour and start being mobilised as proxies for allocation struggle or enforcement reluctance. The show is familiar: rival interpretations of survey coverage, disputed baselines, public claims that advice is biased or incomplete, and growing difficulty in getting parties to accept common numbers even provisionally. The explanation is that science becomes a stress point when its institutional role changes from informing a shared process to supplying ammunition for incompatible political positions. The implication is sobering. Once the numbers themselves are treated as partisan terrain, every later management decision becomes harder to stabilise.

This does not mean contested science should automatically be read as bad faith. A sensible limitation belongs here. Stocks do shift unevenly. Survey design can lag ecological change. Model assumptions can be strained by new productivity patterns or altered spatial distribution. Some scientific disputes reflect real methodological uncertainty rather than strategic obstruction. That is exactly why they are such important warning signs. A governance system that cannot distinguish between legitimate scientific strain and politically useful doubt is already vulnerable. The early warning lies not in disagreement alone, but in whether institutions can still hold disagreement within a trusted process.

Weak sanction pipelines are often the least dramatic warning sign and among the most revealing. A fishery may appear well monitored. Suspicious events may be noticed. Inspections may even increase. Yet if cases fail to mature into credible penalties because documentation chains are weak, agencies overloaded, prosecutors inconsistent, or courts slow, the system begins to teach fleets the wrong lesson. The show is mundane: delayed files, uneven penalties, dropped cases, reliance on warnings where stronger follow-through might once have been expected, and growing separation between what is seen and what becomes consequential. The explanation is that enforcement depends on administrative continuity, not on detection alone. The implication is why weak sanction pipelines are not merely a back-office nuisance. They are structural warnings that visibility is outrunning usable authority.

This is where monitoring limits connect directly to governance design. Authorities often respond to fisheries pressure by demanding more observation: more sensors, more patrols, more data, more cameras, more reporting. Sometimes that is necessary. Yet an early warning framework asks a prior question: where will the additional information go, and what design features will turn it into calm rather than more contest? If a state cannot process surges in evidence, if thresholds for action remain vague, or if verification norms are weak, then improved monitoring may simply illuminate stress more brightly without reducing it. The system will know more and govern no better. That is not readiness. It is exposure.

The quiet motif here is the crack in the harbour wall. At first it looks small, perhaps seasonal, perhaps manageable. No one claims the wall is collapsing. But the crack widens under repeated weather, bad drainage, and deferred repair. Fisheries stress points work the same way. A contested survey, a crowded season, a thin sanction record, a new cluster of vessels in newly important waters - each can be dismissed as an isolated difficulty. Together they reveal where the structure is already carrying more than it was designed to bear.

This is why minor disputes should be read analytically, not just administratively. A boarding dispute may indicate more than a difficult captain. It may reveal that the pressure of a compressed season has outpaced clear rules of contact. A quarrel over distribution may reveal not only national firmness, but the absence of revisable allocation procedures. Public attacks on stock assessment may reveal deeper stress in the legitimacy of science. Repeated weak penalties may reveal that the enforcement pipeline is thinning under the load. To read these signs properly is not to panic at every irregularity. It is to ask what kind of structural weakness the irregularity may be expressing.

The deeper lesson is that conflict prevention begins with better diagnosis of strain. A fisheries regime becomes dangerous not when the first incident occurs, but when its institutions have already lost the ability to treat early warnings as design problems rather than as isolated inconveniences. If warning signs are recognised early, governments can adjust roles, tighten verification, redesign incentives, clarify triggers, or strengthen procedural off-ramps before friction becomes theatre. If they are ignored, the same signs accumulate until a later incident is misdescribed as a surprise.

The unresolved question, then, is not whether the next dispute will offer warning signs. It almost certainly will. The harder question is whether authorities will read those signs as mere operational clutter - one more contested inspection, one more disputed survey, one more delayed case - or as evidence that the system's real vulnerability lies in how ecological pressure, monitoring limits, and institutional design are beginning to line up in ways that make a larger rupture increasingly ordinary.

What a stable frontier requires

In 2018, the Agreement to Prevent Unregulated High Seas Fisheries in the Central Arctic Ocean entered political life with a restraint many observers found almost paradoxical. Here was a frontier opening in ecological and strategic imagination, and the response was not triumphal expansion but a decision to slow things down, institutionalise consultation, and refuse the fantasy that absence of conflict could be secured by pretending nothing was changing. The significance of that move lay not only in precaution. It lay in a different idea of stability. A frontier can remain unsettled without becoming ungovernable.

That is the right place to begin because stability in distant-water fisheries is often described in the wrong register. People reach instinctively for images of fixed borders, settled shares, final agreements, or a legal line sharp enough to stop argument at the moment it appears. Those are understandable desires. They are also poor guides to the conditions under which moving stocks are actually managed. Fish do not respect the emotional appeal of permanence. Institutions that demand permanence from a shifting ecological system eventually become either brittle or theatrical. The map stays still. The pressure does not.

A stable frontier, then, is not one in which change has ceased. It is one in which change does not automatically overwhelm rule. The show is practical: quotas can be revised without making every revision feel like surrender; monitoring can produce records that travel across agencies and borders without dissolving into mutual accusation; incidents at sea can be routed into procedures that cool rather than amplify their political meaning. The name for this is credible order under change. The explanation is simple enough. Stability comes from institutional capacity to absorb movement, not from the wish that movement will stop. The implication is important. A fishery can remain contested, unequal, and politically sensitive while still being stable if the rules for adjustment, verification, and de-escalation are trusted enough to carry strain without breaking.

This is why frozen borders and permanent agreement are fantasies in this domain, not because law is irrelevant, but because law alone cannot make a mobile resource behave as if it belonged to an older settlement indefinitely. The deeper mistake is to imagine that conflict disappears once rights are declared clearly enough. Often the reverse happens. A rigid legal framework can intensify pressure if ecological change makes its distributive effects look increasingly detached from reality. Fleets then test. States harden. Science becomes politicised. Every patrol encounter starts carrying too much constitutional weight. A frontier ruled only by fixity invites the very instability it claims to prevent.

Revisable allocations are one answer to this problem because they make movement governable before it becomes theatrical. The value of revision does not lie in endless fluidity. It lies in bounded adjustment. If parties know in advance how certain shifts in distribution, access, or conditions will be handled, they have less reason to turn every altered season into a contest of nerve. The mackerel disputes in the north-east Atlantic made the cost of not having such mechanisms painfully visible. What might have been processed through a known adaptive design instead became a repeated argument over fairness, standing, and first-mover advantage. The absence of revisability did not preserve order. It forced too much of order to be renegotiated in public under pressure.

And yet revision by itself is not enough. A revisable system without credible verification is simply a more elaborate way to produce mistrust. This is why stable frontiers require institutions that can verify behaviour and ecological change in forms that others are prepared, however grudgingly, to accept. Verification is not the same as surveillance. A state can watch constantly and still fail to govern if what it sees cannot be attributed, audited, or translated into accepted evidence. The show is familiar: disputed tracks, contested monitoring gaps, camera footage challenged on procedural grounds, data shared too late or in incompatible form, sanction cases collapsing because the evidentiary chain is weaker than the public language that preceded it. A stable frontier needs more than sight. It needs records that can survive contact with rivals, courts, agencies, and domestic politics.

This is where norms matter more than machinery. Tools can widen the field of observation, but they do not settle what counts as a reliable trace, a meaningful anomaly, or a legitimate inference. Standards for tamper evidence, auditability, chain of custody, and data sharing are what prevent technology from becoming another frontier of accusation. The issue is not whether anyone will ever try to manipulate, conceal, or reinterpret evidence. They will. The issue is whether the institutional environment makes such efforts detectable and expensive enough that the evidentiary order remains intact. The seal matters.

De-escalatory procedures form the third requirement because even well-designed allocation rules and good verification systems cannot remove every incident. Vessels will still crowd boundaries, seasons will still compress, inspections will still be contested, and domestic audiences will still demand sharp language at awkward moments. Stability therefore depends on whether the system provides off-ramps before every minor collision of interests becomes a test of national resolve. Hotlines, technical channels, interim arrangements, agreed procedures for document exchange, and ordinary habits of communication are often dismissed as mere process. In reality they are among the most important security devices in distant-water governance. They move quarrels from decks and headlines into timelines and files. That movement is often the difference between tension and crisis.

A useful counter-example should be admitted. Stability is not always produced by more flexibility, more procedure, or more consultation. Some arrangements become so adjustable that no one believes anything is settled enough to invest in. Some procedural systems become so elaborate that they provide cover for delay rather than routes out of conflict. Some states invoke verification standards selectively, trusting process when it favours them and denouncing it when it does not. This matters because adaptive governance can fail by becoming vague, slow, or strategically manipulable. The answer is not maximal softness. It is disciplined adaptability: rules that can move, but not everywhere; verification that can be challenged, but not endlessly; procedures that can buy time, but not become substitutes for decision.

This is why the frontier remains. It cannot be legislated away, morally pacified, or technologically neutralised. Stocks will continue to move, governments will continue to protect domestic constituencies, and fleets will continue to look for profitable margins in any system that leaves them available. The point of governance is therefore not to abolish frontier politics, but to civilise it institutionally. A stable frontier is one in which states still disagree, but do so within arrangements strong enough that disagreement does not constantly spill into unmanaged confrontation. It is one in which rules remain credible even when they are revised, in which evidence remains usable even when it is contested, and in which procedures remain trusted enough that delay can cool rather than corrode.

The practical test of such stability is not peace in the sentimental sense. It is whether a difficult season produces manageable friction or cumulative institutional damage. When a stock shifts, can the system adjust without reopening every sovereignty claim? When a vessel behaves ambiguously, can the evidence be handled without the whole monitoring order becoming suspect? When an inspection goes wrong, can the incident be contained procedurally before public narrative hardens into national grievance? These are not abstract questions. They are the ordinary examinations by which distant-water governance proves whether it has moved beyond symbolic control and into something closer to durable order.

The quiet motif here is the hinge. A frontier does not stop being a frontier because the door is removed. It becomes governable because the hinge holds while the door moves. Too rigid, and it snaps under pressure. Too loose, and it fails to close at all. The craft lies in building a hinge that can bear repeated motion without pretending motion is a design flaw. That is what revisable allocations, credible verification, and de-escalatory procedure amount to in combination. They are not final answers. They are load-bearing parts.

What is feasible, then, is narrower than the rhetoric of absolute control and broader than the fatalism of inevitable conflict. It is feasible to design institutions that adapt without dissolving, that observe without overclaiming, and that cool disputes without requiring affection or permanent agreement. It is not feasible to expect distant-water fisheries to rest on frozen distributions, complete visibility, or a final settlement after which political bargaining no longer matters. A serious governance theory begins by separating those two categories. Fantasy flatters. Institutions endure.

The harder question, then, is not whether the frontier can be made safe by stronger language, better sensors, or one more definitive legal settlement. It is whether the institutions surrounding that frontier are credible enough under change that the next shift in fish, effort, or pressure is treated as something to be governed rather than as fresh proof that only rigidity or unilateralism can still deliver order.

About the Author

Maren Ostergaard is a nonfiction writer and analyst concerned with how institutions cope when environmental change collides with economic incentives and political pride. Her work approaches the Arctic not as an empty periphery but as a densely governed space where maps, markets, and measurement systems constantly negotiate with one another. She writes with an editorial commitment to clarity over jargon, and to describing what rules do in practice as well as what they claim on paper.

Ostergaard is especially interested in the gap between formal agreement and operational reality: how monitoring systems produce knowledge and blind spots; how quotas become political settlements; and how routine inspections can carry diplomatic meaning. Across her work runs a historical sensibility shaped by the long North Atlantic record of seafaring, seasonal work, and contested access to living resources. That background thread matters because it keeps the analysis grounded in continuity as well as change: today's satellite tracking and data audits sit atop older habits of boundary-testing, bargaining, and reputation.

In The Fisheries Frontier, she brings together ecology-aware governance analysis with an eye for incentives and institutional design. The result is a book written for readers who want to understand how conflict is prevented in practice, and why it sometimes is not, without reducing complex disputes to morality plays or headline drama.

www.ingramcontent.com/pod-product-compliance
Lightning Source LLC
Chambersburg PA
CBHW051355240926
51852CB00001B/34
9789377940058